Engineering Properties
of Soils and Rocks

Engineering Properties of Soils and Rocks

F. G. BELL

Butterworths
London Boston Sydney Durban Wellington Toronto

The Butterworth Group

United Kingdom   **Butterworth & Co (Publishers) Ltd.**
London: 88 Kingsway, WC2B 6AB

Australia   **Butterworths Pty Ltd**
Sydney: 586 Pacific Highway, Chatswood, NSW 2067
Also at Melbourne, Brisbane, Adelaide and Perth

Canada   **Butterworth & Co. (Canada) Ltd**
Toronto: 2265 Midland Avenue, Scarborough,
Ontario M1P 4S1

New Zealand   **Butterworths of New Zealand Ltd**
Wellington: T & W Young Building, 77–85 Customhouse Quay,
1, CPO 472

South Africa   **Butterworth & Co (South Africa) (Pty) Ltd**
Durban: 152–154 Gale Street

USA   **Butterworth (Publishers) Inc**
Boston: 10 Tower Office Park, Woburn, Mass. 01801

First published 1981

© Butterworth & Co. (Publishers) Ltd, 1981

British Library Cataloguing in Publication Data
Bell, Frederick Gladstone
Engineering properties of soils and rocks.
1. Soil mechanics   2. Rock mechanics
I. Title
624'.1513   TA710   80–41145
ISBN 0-408-00537-8

Typeset by Scribe Design, Gillingham, Kent
Printed in England by Billing & Sons Limited,
Guildford, London and Worcester

Preface

The engineering geologist, civil engineer and mining engineer all require a working knowledge of the engineering behaviour of soils and rocks. However, most textbooks of soil mechanics and rock mechanics generally do not provide sufficient data relating to the actual values of the various engineering properties of individual soil or rock types respectively. This therefore was the main reason for the production of this small text, which gives a brief survey of the engineering properties of the major types of soil and rock. It also attempts to show how these properties vary according to their geological setting, the most obvious illustration being the influence of discontinuities and weathering on the engineering performance of rock masses. Similarly, sorting, packing and grain shape all influence the engineering properties of soil, and the mineral composition is of particular importance in the case of deposits of clay.

With this in mind, it is hoped that the book will complement the standard texts now available on soil and rock mechanics and that the student can read one alongside the other. As can be inferred from the above, the book is aimed primarily at undergraduate students in engineering geology, civil engineering and mining engineering. But it is also hoped that it may be of use to the practising engineer; if he is aware of everything contained therein, then reading the text will at least provide an exercise in critical indulgence!

This book is derived in large part from one of the author's previous works, with some updating and with the addition of some new material. It has always been the author's policy to offer a plentiful supply of references, to give maximum facility to those who are interested to pursue the subject in more depth.

Finally the author wishes to thank John F. Bell, who willingly checked the manuscript and proofs, and Don Goodsell, commissioning editor, whose help, as always, proved invaluable.

F.G. Bell
Blyth, Notts

Contents

Chapter 1

Soil Classification

1.1 ORIGIN OF SOIL

Soil is an unconsolidated assemblage of solid particles which may or may not contain organic matter, the voids between the particles being occupied by air and/or water. It derives mainly from rock material broken down by physical or chemical weathering* and by water, wind and ice which are responsible for the varying amounts of transport which particles may undergo prior to deposition. Gravity likewise has a hand in particle transport. The type of breakdown and amount of transport have a significant influence on the character of a deposit (Table 1.1), as has the parental material. In addition changes occur in a deposit after it has been laid down. Accordingly different types of soil evolve, with different grain size distributions, with differing degrees of sorting and packing, and with differently shaped particles.

1.2 BASIS OF CLASSIFICATION : GRAIN SIZE

Classification of soils is made on a basis of certain fundamental properties and provides an ordered framework for their systematic description. Such a classification ideally should indicate the engineering performance of the soil type and should provide a quick means of identification.

The fundamental property upon which most engineering classifications of soils are based is particle size distribution, since it is readily measurable and has an important influence on soil behaviour. Boulders, cobbles, gravels, sands, silts and clays are distinguished as individual groups, each group being given the following symbol and size range:
  1. Boulders (B), over 200 mm
  2. Cobbles (Cb), 60 to 200 mm
  3. Gravel (G), 2 to 60 mm (20 to 60 mm, coarse; 6 to 20 mm, medium; 2 to 6 mm, fine).
  4. Sand (S), 0.06 to 2 mm (0.6 to 2 mm, coarse; 0.2 to 0.6 mm, medium; 0.06 to 0.2 mm, fine)

*Peat, being an organic soil, does not fall within this context

1

5. Silt (M), 0.002 to 0.06 mm (0.02 to 0.06 mm, coarse; 0.006 to 0.02 mm, medium; 0.002 to 0.006 mm, fine)

6. Clay (C), less than 0.002 mm.

The gravel, sand and silt size ranges are further divided up into coarse, medium and fine categories. Sands and gravels are granular materials ideally possessing no cohesion whereas silts and clays are cohesive materials.

Table 1.1 EFFECTS OF TRANSPORTATION ON SEDIMENTS

	Gravity	*Ice*	*Water*	*Air*
Size	Various	Varies from clay to boulders	Various sizes from boulder gravel to muds	Sand size and less
Sorting	Unsorted	Generally unsorted	Sorting takes place both laterally and vertically. Marine deposits often uniformly sorted. River deposits may be well sorted	Uniformly sorted
Shape	Angular	Angular	From angular to well rounded polished surface	Well rounded
Surface texture	Striated surfaces	Striated surfaces	Gravel: rugose surfaces. Sand: smooth, polished surfaces. Silt: little effect	Impact produces frosted surfaces

1.3 BASIS OF CLASSIFICATION : PLASTICITY

These major groups are divided into subgroups on a basis of grading and the plasticity of the fine material (Table 1.2). In the proposed British Soil Classification[1], which is based upon the Casagrande classification[2], granular soils are described as well graded (W) or poorly graded (P). Two further types of poorly graded granular soils are recognised, namely, uniformly graded (Pu) and gap-graded (Pg). Silts and clays are generally subdivided according to their liquid limits (LL) into low (LL = less than 30. L), medium (LL = 35 to 50. M) and high (LL = over 50. H) subgroups. Very high (LL = 70 to 90. V) and extremely high (LL = over 90. E) categories have also been recognised. Each subgroup is given a combined symbol in which the letter describing the predominant size fraction is written first (e.g. GW = well graded gravels; CH = clay with high liquid limit). Silty soil has a restricted plastic range in relation to its liquid limit whilst clay is fully plastic. The plasticity chart for use in soil classification (Table 1.2c) was introduced by Casagrande[2]. Silts tend to plot below, and clays above the A-line. Organic soils also plot below the A-line.

Any group may be referred to as organic if it contains a significant proportion of organic matter, in which case the letter O is suffixed to the group symbol (e.g. CVSO = organic clay of very high liquid limit with sand). The symbol Pt is given to peat.

1.4 MIXED SOILS

In many soil classifications boulders and cobbles are removed before an attempt is made at classification, for example, their proportions are recorded separately in the proposed British Soil Classification. Their presence should be recorded in the soil description, a plus sign being used in symbols for soil mixtures, for example, G + Cb for gravel with cobbles. The British Soil Classification has proposed that very coarse deposits should be classified as follows:

1. BOULDERS	Over half of the very coarse material is of boulder size (over 200 mm). May be described as cobbly BOULDERS if cobbles are an important second constituent in the very coarse fraction.
2. COBBLES	Over half of the very coarse material is of cobble size (200–60 mm). May be described as bouldery COBBLES if boulders are an important second constituent in the very coarse fraction.

Mixtures of very coarse material and soil can be described by combining the terms for the very coarse constituent and the soil constituent as follows:

1. clean BOULDERS/COBBLES , up to 5% soil
2. BOULDERS/COBBLES with some SOIL * 5–20% soil
3. BOULDERS/COBBLES with much SOIL* 20–50% soil
4. SOIL* with many BOULDERS/COBBLES 50–20% boulders
5. SOIL* with some BOULDERS/COBBLES 20–5% boulders
6. SOIL* with occasional BOULDERS/COBBLES up to 5% boulders

*Give soil name in brackets, e.g. Cobbly BOULDERS with some SOIL (sand with some fines)

Mixed soil types can be indicated as follows:

1. sandy GRAVEL gravelly SAND	With sand sized or gravel sized material as an important second constituent of the coarse fraction
2. clean GRAVEL/SAND	With under 5% fines (G; S)
3. GRAVEL/SAND with some fines (F)	With 5 to 20% fines (can be distinguished as silt or clay: G-F, G-M, G-C; S-F, S-M, S-C)
4. GRAVEL/SAND with much fines	With 20 to 50% fines (can be distinguished as silt or clay: GF, GM, GC; SF, SM, SC)
5. SILT/CLAY with gravel or with sand	With 30–50% gravel or sand (MG, MS; CG, CS)

The British Soil Classification also has proposed that classification can be made either by rapid assessment or by full laboratory procedure (Tables 1.2a and 1.2b). It was recommended that if classification was by means of rapid assessment, brackets should enclose the group symbol indicating a lower degree of accuracy, whereas if full laboratory procedure was used for classification the group symbol should not be bracketed. For a comparison with the Unified Soil Classification which is used in the United States see Table 1.2c. A note on the engineering uses of these soils is provided in Table 1.2d.

References

1. Code Drafting Committee, 'Proposed British soil classification system for use in the revised Code of Practice on site investigations. *Inst. Civ. Engrs.* (1976).
2. Casagrande, A. 'Classification and identification of soils, *Trans. A.S.C.E.*, **113**, 901–992 (1948).
3. Wagner, A.A., 'The use of the Unified Soil Classification System for the Bureau of Reclamation, *Proc. 4th Int. Conf. Soil Mech. Found. Engng., London*, **1**, 125–134 (1957).

Table 1.2a THE BRITISH SOIL CLASSIFICATION SYSTEM FOR ENGINEERING PURPOSES (BSCS). FIELD OR RAPID IDENTIFICATION
[First remove material coarser than 60 mm and record as *COBBLES* (60–200 mm) or *BOULDERS* (over 200 mm)]

SOIL GROUPS		SYMBOL		TEXTURE	Remoulded at suitable moisture content	
					COHESION Ability to stick together	PLASTICITY Ability to deform without rupture
GRAVEL and SAND may be qualified sandy GRAVEL and gravelly SAND where appropriate						
COARSE SOILS More than half coarser than 60 μm (i.e. visible to eye or gritty to feel)	GRAVELS More than half of coarse material is of gravel size (coarser than 2 mm)					
	Clean *GRAVEL*	G	GW	Little or no fines. GW has wide range of grain sizes, well distributed. GP has one size predominating, GPu, or missing, GPg	None	None
			GP			
	GRAVEL with some silt	G-F	G-M	Gravel with some silt	None to low	None
	GRAVEL with some clay		G-C	Gravel with some clay	Low to medium	None
	GRAVEL with much silt	GF	GM	Gravel with much silt	None to low	None
	GRAVEL with much clay		GC	Gravel with much clay	Medium to high	Low to medium
	SANDS More than half of coarse material is of sand size (finer than 2 mm)					
	Clean *SAND*	S	SW	Little or no fines. SW has wide range of grain sizes, well distributed. SP has one size predominating, SPu, or missing SPg	Low	None
			SP			
	SAND with some silt	S-F	S-M	Sand with some silt	Low	None
	SAND with some clay		S-C	Sand with some clay	Low to medium	None to low
	SAND with much silt	SF	SM	Sand with much silt	Low to medium	Low to medium
	SAND with much clay		SC	Sand with much clay	Medium to very high	Medium to very high

Category	Group	Sub-group	Symbol	Group symbol	Field identification	DRY STRENGTH Dried from plastic limit	TOUGHNESS Consistency at plastic limit	DILATANCY
FINE SOILS less than half coarser than 60 μm (i.e. visible to eye or gritty to feel)	SILTS and CLAYS with gravel or sand 50–70% fines	SILT with gravel / CLAY with gravel	MG / CG	FC	Coarse material mainly over 2 mm			
		SILT with sand / CLAY with sand	MS / CS	FS	Coarse material mainly under 2 mm			
					Similar to SILT and CLAY groups (see below), but with properties modified by a considerable proportion of coarse material			
					First remove coarser particles by hand			
	SILTS and CLAYS 70–100% fine (finer than 60 μ)	SILT / CLAY	M / CI					
		SILT of low liquid limit	ML		Dries moderately quickly and can be brushed off the fingers. Inorganic	None to low	None to low	High to moderate
		CLAY of low liquid limit	CL			Medium	Medium	Low
		of intermediate liquid limit	CI		Sticks to fingers and dries slowly.	High	Medium	None
		of high liquid limit	CH			High	High	None
		of very high liquid limit	CV		Shrinks appreciably on drying, usually showing cracks	Very high	High	None
ORGANIC SOILS	'Organic' prefixed to any group name and letter O, suffixed to symbol, e.g. CO				Organic matter suspected to be a significant constituent. Dark colour, distinctive odour, moisture content may be very high			
PEAT	Pt				Peat soil consists predominantly of plant resins, which may be fibrous or amorphous. Dark colour, distinctive odour, low bulk density, moisture content may be very high			

Table 1.2b BRITISH SOIL CLASSIFICATION SYSTEM FOR ENGINEERING PURPOSES (BSCS)

[First remove material coarser than 60 mm and record as COBBLES (60 mm–200 mm) or BOULDERS (over 200 mm)]

Soil Groups		Sub-Groups and laboratory identification			
		Group Symbol	Sub-Group Symbol	Fines % less than 60 μm	Sub-Group name
GRAVEL and *SAND* may be qualified *Sandy GRAVEL* and *Gravelly SAND* where appropriate					
COARSE SOILS More than 50% coarser than 60 μm	*GRAVELS* More than 50% of coarse material is of gravel size (coarser than 2 mm)				
	Clean *GRAVEL*	G GW	GW	0–5	Well-graded
		GP	GP GPu GPg		Poorly/uniformly/gap graded } clean *GRAVEL*
	GRAVEL with some silt	G-F G-M	GWM GPM	5–20	Well/poorly graded *GRAVEL* with some silt
	GRAVEL with some clay	G-C	GWC GPC		clay
	GRAVEL with much silt	GF GM	GM	20–50	*GRAVEL* with much silt: sub-divide like GC
	GRAVEL with much clay	GC	GCL GCI GCH GCV		*GRAVEL* with much clay of low/intermediate/high/very high plasticity
	SANDS More than 50% of coarse material is of sand size (finer than 2 mm)				
	Clean *SAND*	S SW	SW	0–5	Well graded
		SP	SP SPu SPg		Poorly/uniformly/gap graded } clean *SAND*
	SAND with some silt	S-F S-M	SWM SPM	5–20	Well/poorly graded *SAND* with some silt
	SAND with some clay	S-C	SWC SPC		clay
	SAND with much silt	SF SM	SM	20–50	*SAND* with much silt: sub-divide like SC
	SAND with much clay	SC	SCL SCI SCH SCV		*SAND* with much clay of low/intermediate/high/very high plasticity

FINE SOILS / SILTS and CLAYS					Liquid Limit %	
FINE SOILS more than 50% finer than 60 μm	**SILTS and CLAYS** with gravel or sand 50–70% fines	*SILT* with gravel	FG	MG	MG		*SILT* (divide like CG) with gravel
		CLAY with gravel		CG	CLG	<35	*CLAY* of low liquid limit with gravel
					CIG	35–50	of intermediate liquid limit with gravel
					CHG	50–70	of high liquid limit with gravel
					CVG	70–90	of very high liquid limit with gravel
		SILT with sand	FS	MS	MS		*SILT* with sand
		CLAY with sand		CS	CLS etc		*CLAY* with sand sub-divide like GC
	SILTS and CLAYS 70–100% fines	*SILT*	F	M	M		*SILT*: sub-divide like C
		CLAY		C	CL	<35	*CLAY* of low liquid limit
					CI	35–50	of intermediate liquid limit
					CH	50–70	of high liquid limit
					CV	70–90	of very high liquid limit

ORGANIC SOILS — Descriptive letter O suffixed to any group or sub-group symbol — Organic matter suspected to be a significant constituent. Example: *MHO, organic SILT of high liquid limit*

PEAT — Pt — Peat soils consist predominantly of plant remains, which may be fibrous or amorphous.

Table 1.2c UNIFIED SOIL CLASSIFICATION (After Wagner, 1957)[3]

Field identification procedures (excluding particles larger than 76 mm, and basing fractions on estimated weights)			Group Symbols	Typical names	Information required for describing soils	Laboratory classification criteria
Coarse-grained soils More than half of material is *larger* than No. 200 sieve size[b]	Gravels More than half of coarse fraction is *larger* than No. 7 sieve size*	Clean gravels (little or no fines)		GW	Well graded gravels, gravel-sand mixtures, little or no fine	Give typical name; indicate approximate percentages of sand and gravel; maximum size; angularity, surface condition, and hardness of the coarse grains; local or geologic name and other pertinent descriptive information: and symbols in parentheses	Determine percentages of gravel and sand from grain size curve. Depending on percentage of fines (fraction smaller than No. 200 sieve size) coarse grained soils are classified as follows: Less than 5%: *GW, GP, SW, SP.* More than 12%: *GM, GC, SM, SC.* 5% to 12%: Borderline cases requiring use of dual symbols	$C_u = \dfrac{D_{60}}{D_{10}}$ Greater than 4 $C_c = \dfrac{(D_{30})^2}{D_{10} \times D_{60}}$ Between 1 & 3
				GP	Poorly graded gravels, gravel-sand mixtures, little or no fine			Not meeting all gradation requirements for *GW*
		Gravels with fines (appreciable amount of fines)	Nonplastic fines (for identification procedures see *ML* below)	GM	Silty gravels, poorly graded gravel-sand-silt mixtures	For undisturbed soils add information on stratification, degree of compactness, cementation, moisture conditions and drainage characteristics		Atterberg limits below "A" line or *PI* less than 4 · Atterberg limits above "A" line with *PI* greater than 7 · Above "A" line with *PI* between 4 and 7 are *borderline* cases requiring use of dual symbols
			Plastic fines (for identification procedures, see *CL* below)	GC	Clayey gravels, poorly graded gravel-sand-clay mixtures			
	Sands More than half of coarse fraction is *smaller* than No. 7 sieve size*	Clean sands (little or no fines)	Wide range in grain sizes and substantial amounts of all intermediate particle sizes	SW	Well graded sands, gravelly sands, little or no fines	Example: *Silty sand*, gravelly; about 20% hard, angular gravel particles 12.5 mm maximum size; rounded and subangular sand grains coarse to fine, about 15% nonplastic fines with low dry strength; well compacted and moist in place; alluvial sand; *(SM)*	Use grain size curve in identifying the fractions as given under field identification	$C_u = \dfrac{D_{60}}{D_{10}}$ Greater than 6 $C_c = \dfrac{(D_{30})^2}{D_{10} \times D_{60}}$ Between 1 & 3
			Predominantly one size or a range of sizes with some intermediate sizes missing	SP	Poorly graded sands, gravelly sands, little or no fines			Not meeting all gradation requirement for *SW*
		Sands with fines (appreciable amount of fines)	Nonplastic fines (for identification procedures, see *ML* below)	SM	Silty sands, poorly graded sand-silt mixtures			Atterberg limits below 'A' line or *PI* less than 4 · Atterberg limits below 'A' line with *PI* greater than 7 · Above "A" line with *PI* between 4 and 7 are *borderline* cases requiring use of dual symbols
			Plastic fines (for identification procedures, see *CL* below	SC	Clayey sands, poorly graded sand-clay mixtures			

(For the two rows with "Clean gravels" — Wide range in grain size and substantial amounts of all intermediate particle sizes (GW); Predominantly one size or a range of sizes with some intermediate sizes missing (GP).)

Plasticity chart for laboratory classification of fine grained soils

Use grain size curve in identifying the functions as given under field identification

Fine-grained soils — More than half of material is smaller than No. 200 sieve size b	Identification Procedures on Fraction smaller than No. 40 Sieve Size			Symbol	Typical names	
	DRY STRENGTH (crushing characteristics)	DILATANCY (reaction to shaking)	TOUGHNESS (consistency near plastic limit)			Give typical name; indicate degree and character of plasticity, amount and maximum size of coarse grains; colour in wet condition, odour if any, local or geologic name, and other pertinent descriptive information, and symbol in parentheses. For undisturbed soils add information on structure, stratification, consistency in undisturbed and remoulded states, moisture and drainage conditions. Example: *Clayey silt*, brown: slightly plastic; small percentage of fine sand; numerous vertical root holes; firm and dry in place; loess; (*ML*)
Silts and clays liquid limit less than 50	None to slight	Quick to slow	None	*ML*	Inorganic silts and very fine sands, rock flour, silty or clayey fine sands with slight plasticity	
	Medium to high	None to very slow	Medium	*CL*	Inorganic clays of low to medium plasticity, gravelly clays, sandy clays, silty clays, lean clays	
	Slight to medium	Slow	Slight	*OL*	Organic silts & organic silt-clays of low plasticity	
Silts and clays liquid limit greater than 50	Slight to medium	Slow to none	Slight to medium	*MH*	Inorganic silts, micaceous or diatomaceous fine sandy or silty soils, elastic silts	
	High to very high	None	High	*CH*	Inorganic clays of high plasticity, fat clays	
	Medium to high	None to very slow medium	Slight to medium	*OH*	Organic clays of medium to high plasticity	
Highly organic soils	Readily identified by colour, odour, spongy feel and frequently by fibrous texture			*Pt*	Peat and other highly organic soils	

Footnotes to Table 1.2c

aBoundary classifications. Soils possessing characteristics of two groups are designated by combinations of group symbols. For example *GW–GC*, well graded gravel-sand mixture with clay binder.

bAll sieve sizes on this chart are US standard. The No 200 sieve size is about the smallest particle visible to the naked eye.

*For visual classification, the ¼ in size may be used as equivalent to the No 7 sieve size.

Field Identification Procedure for Fine Grained Soils or Fractions

These procedures are to be performed on the minus No. 40 sieve size particles, approximately 1/64 in. For field classification purposes, screening is not intended, simply remove by hand the coarse particles that interfere with the tests.

Dilatancy (reaction to shaking):

After removing particles larger than No. 40 sieve size, prepare a pat of moist soil with a volume of about one-half cubic inch. Add enough water if necessary to make the soil soft but not sticky.

Place the pat in the open palm of one hand and shake horizontally, striking vigorously against the other hand several times. A positive reaction consists of the appearance of water on the surface of the pat which changes to a livery consistency and becomes glossy. When the sample is squeezed between the fingers, the water and gloss disappear from the surface, the pat stiffens and finally it cracks and crumbles. The rapidity of appearance of water during shaking and of its disappearance during squeezing assist in identifying the character of the fines in a soil.

Very fine clean sands give the quickest and most distinct reaction whereas a plastic clay has no reaction. Inorganic silts, such as a typical rock flour, show a moderately quick reaction.

Dry Strength (Crushing characteristics):

After removing particles larger than No. 40 sieve size, mould a part of soil to the consistency of putty, adding water if necessary. Allow the pat to dry completely by oven, sun or air drying, and then test its strength by breaking and crumbling between the fingers. This strength is a measure of the character and quantity of the colloidal fraction contained in the soil. The dry strength increases with increasing plasticity.

High dry strength is characteristic for clays of the CH group. A typical inorganic silt possesses only very slight dry strength. Silty fine sands and silts have about the same slight dry strength, but can be distinguished by the feel when powdering the dried specimen. Fine sand feels gritty whereas a typical silt has the smooth feel of flour.

Toughness (Consistency near plastic limit):

After removing particles larger than the No. 40 sieve size, a specimen of soil about one-half inch cube in size, is moulded to the consistency of putty. If too dry, water must be added and if sticky, the specimen should be spread out in a thin layer and allowed to lose some moisture by evaporation. Then the specimen is rolled out by hand on a smooth surface or between the palms into a thread about one-eighth inch in diameter. The thread is then folded and re-rolled repeatedly. During this manipulation the moisture content is gradually reduced and the specimen stiffens, finally loses its plasticity, and crumbles when the plastic limit is reached.

After the thread crumbles, the pieces should be lumped together and a slight kneading action continued until the lump crumbles.

The tougher the thread near the plastic limit and the stiffer the lump when it finally crumbles, the more potent is the colloidal clay fraction in the soil. Weakness of the thread at the plastic limit and quick loss of coherence of the lump below the plastic limit indicate either inorganic clay of low plasticity, or materials such as kaolin-type clays and organic clays which occur below the A-line.

Highly organic clays have a very weak and spongy feel at the plastic limit.

Table 1.2d ENGINEERING USE CHART (after Wagner, 1957)[3]

Typical names of soil groups	Group symbols	Important properties — Permeability when compacted	Shearing strength when compacted and saturated	Compressibility when compacted and saturated	Workability as a construction material	Rolled earth dams — Homogeneous embankment	Core	Shell	Canal sections — Erosion resistance	Compacted earth lining	Foundations — Seepage important	Seepage not important	Roadways Fills — Frost heave not possible	Frost heave possible	Surfacing
Well-graded gravels, gravel-sand mixtures, little or no fines	GW	pervious	excellent	negligible	excellent	—	—	1	1	—	—	1	1	1	1
Poorly graded gravels, gravel-sand mixtures, little or no fines	GP	very pervious	good	negligible	good	—	—	2	2	—	—	3	3	3	1
Silty gravels, poorly graded gravel-sand-silt mixtures	GM	semi-pervious to impervious	good	negligible	good	2	4	—	4	4	1	4	4	9	5
Clayey gravels, poorly graded gravel-sand-clay mixtures	GC	impervious	good to fair	very low	good	1	1	—	3	1	2	6	5	5	1
Well-graded sands, gravelly sands, little or no fines	SW	pervious	excellent	negligible	excellent	—	—	3 if gravelly	6	—	—	2	2	2	4
Poorly graded sands, gravelly sands, little or no fines	SP	pervious	good	very low	fair	—	—	4 if gravelly	7 if gravelly	—	—	5	6	4	—

Relative desirability for various uses (Graded from 1 (highest) to 14 (lowest))

Silty sands, poorly graded sand-silt mixtures	SM	semi-pervious to impervious	good	low	fair	4	5 erosion critical	–	8 if gravelly	5 erosion critical	3	7	8	10	6
Clayey sands, poorly graded sand-clay mixtures	SC	impervious	good to fair	low	good	5	2	–	3	2	4	8	7	6	2
Inorganic silts and very fine sands, rock flour, silty or clayey fine sands with slight plasticity	ML	semi-pervious to impervious	fair	medium	fair	6 erosion critical	6 erosion critical	–	6 erosion critical	6	6	9	10	11	–
Inorganic clays of low to medium plasticity, gravelly clays, sandy clays, silty clays, lean clays	CL	impervious	fair	medium	good to fair	9	3	–	5	3	5	10	9	7	7
Organic silts and organic silt-clays of low plasticity	OL	semi-pervious to impervious	poor	medium	fair	7 erosion critical	7 erosion critical	–	8	7	7	11	11	12	–
Inorganic silts, micaceous or diatomaceous fine sandy or silty soils, elastic silts	MH	semi-pervious to impervious	fair to poor	high	poor	–	–	–	9	–	8	12	12	13	–
Inorganic clays of high plasticity, fat clays	CH	impervious	poor	high	poor	10	8 volume change critical	–	7	9	9	13	13	8	–
Organic clays of medium to high plasticity	OH	impervious	poor	high	poor	–	–	–	10	10	10	14	14	14	–
Peat and other highly organic soils	Pt	–	–	–	–	–	–	–	–	–	–	–	–	–	–

Chapter 2

Coarse Grained Soils

The composition of a gravel deposit reflects not only the source rocks of the area from which it was derived but is also influenced by the agents responsible for its formation and the climatic règime in which it was or is being deposited. The latter two factors have a varying tendency to reduce the proportion of unstable material. Relief also influences the nature of a gravel deposit, for example, gravel production under low relief is small and the pebbles tend to be chemically inert residues such as vein quartz, quartzite, chert and flint. By contrast high relief and rapid erosion yield coarse, immature gravels.

Sands consist of a loose mixture of mineral grains and rock fragments. Generally they tend to be dominated by a few minerals, the chief of which is quartz. There is a presumed dearth of material in those grades transitional to gravel on the one hand and silt on the other (see Glossop and Skempton[1]). Sands vary appreciably in their textural maturity.

2.1 SOIL FABRIC

The engineering behaviour of a soil is a function of its structure or fabric, which in turn is a result of the geological conditions governing deposition and the subsequent stress history. The macro-structure of a soil includes its bedding, laminations, fissures, joints and tension cracks, all of which can exert a dominant influence on the shear strength and drainage characteristics of a soil mass. The micro-structure of a sand or gravel refers to its particle arrangement which in turn involves the concept of packing — in other words the spacial density of particles in the aggregate (see Kahn[2]).

The conceptual treatment of packing begins with a consideration of the arrangement of spherical particles of equal size. These can be packed either in a disorderly or systematic fashion. The closest type of systematic packing is rhombohedral packing whereas the most open type is cubic packing, the porosities approximating to 26% and 48% respectively. Put another way, the void ratio of a well sorted and perfectly cohesionless aggregate of equidimensional grains can range between extreme values of about 0.35 and 1.00. If the void ratio is more than unity the micro-structure will be collapsable or metastable. If a large number of spheres of equal size is arranged in any systematic packing pattern then there is a certain diameter ratio for smaller spheres which can just

pass through the throats between the larger spheres into the interstices, for
example, in rhombohedral packing this critical diameter is 0.154 D (D being the
diameter of the larger spheres). However, a considerable amount of disorder
occurs in most coarse grained deposits and, according to Graton and Frazer[3],
there are colonies of tighter and looser packing within any deposit.

In a single grain structure individual particles are bulky and pore passages
have average diameters of the same order of magnitude as smaller particle
diameters. There is virtually no effective combination of particles to form
aggregates. Each particle functions individually in the soil framework, and
particles are in contact with one another, so that the movement of any individual grain is influenced by the position of adjacent grains. For most equilibrium conditions in coarse grained soil the soil framework serves exclusively as
the stressed member.

Size and sorting have a significant influence on the engineering behaviour of
granular soils; generally speaking the larger the particles, the higher the strength.
Deposits consisting of a mixture of different sized particles are usually stronger
than those which are uniformly graded. However, the mechanical properties of
such sediments depends mainly on their density index (formerly relative density)
which in turn depends on packing. For instance, densely packed sands are almost
incompressible whereas loosely packed deposits, located above the water table,
are relatively compressible but otherwise stable. If the density index of a sand
varies erratically this can give rise to differential settlement. Generally settlement
is relatively rapid. However, when the stresses are large enough to produce
appreciable grain fracturing, there is a significant time lag.

Greater settlement is likely to be experienced in granular soils where foundation level is below the water table rather than above. Additional settlement
may occur if the water table fluctuates or the ground is subject to vibrations.
Although the density index may decrease in a general manner with decreasing
grain size there is ample evidence to show, for example, that water deposited
sands with similar grain size can vary between wide limits. Hence factors other
than grain size, such as rate of deposition and particle shape, influence the
density index.

2.2 DEFORMATION OF GRANULAR SOIL

Two basic mechanisms contribute towards the deformation of granular soil,
namely, distortion of the particles, and the relative motion between them.
These mechanisms are usually interdependent. At any instant during the deformation process different mechanisms may be acting in different parts of the soil
and these may change as deformation continues. Interparticle sliding can occur
at all stress levels, the stress required for its initiation increasing with initial stress
and decreasing void ratio. Crushing and fracturing of particles begins in a minor
way at small stresses, becoming increasingly important when some critical stress
is reached. This critical stress is smallest when the soil is loosely packed and
uniformly graded, and consists of large, angular particles with a low strength.
Usually fracturing becomes important only when the stress level exceeds 3.5
MPa.

The internal shearing resistance of a granular soil is generated by friction
developed when grains in the zone of shearing are caused to slide, roll and rotate
against each other. At the commencement of shearing in a sand some grains are

moved into new positions with little difficulty. The normal stress acting in the direction of movement is small but eventually these grains occupy positions in which further sliding is more difficult. By contrast other grains are so arranged in relation to the grains around them that sliding is difficult. They are moved without sliding by the movements of other grains. The frictional resistance of the former is developed as the grains become impeded whereas in the latter case it is developed immediately. The resistance to rolling represents the sum of the behaviour of all the particles, and the resistance to sliding is essentially attributable to friction which, in turn, is proportional to the confining stress.

Frictional resistance is built up gradually and consists of establishing normal stresses in the intergranular structure as the grains push or slide along (see Cornforth[4]). At the same time sliding allows the structure to loosen in dilatant soils which reduces normal stress. The maximum shearing resistance is a function of these two factors. The packing and external stress conditions govern the amount of sliding by individual grains in mobilizing shearing resistance. According to Cornforth the latter factor is the more important and in fact is really a strain condition. He therefore concluded that the strain condition during shear is a major factor contributing to the strength of sand.

2.3 STRENGTH AND DISTORTION

Interlocking grains contribute a large proportion of the strength in densely packed granular soils and shear failure occurs by overcoming the frictional resistance at the grain contacts. Conversely, interlocking has little or no effect on the strength of very loosely packed coarse grained soils in which the mobility of the grains is greater (see Borowicka[5]). Figure 2.1 shows that dense sand has a high peak strength and that, when it is subjected to shear stress, it expands up to the point of failure, after which a slight decrease in volume may occur. Loose sand, on the other hand, compacts under shearing stress and its residual strength may be similar to that of dense sand, and tends to remain so. Hence a constant void ratio is obtained, that is, the critical volume condition which has a critical angle of friction and a critical void ratio. These are independent of initial density being a function of the normal effective stress at which shearing occurs. For a discussion of the shear strength of sands see Rowe[6] and Frossard[7].

Both curves in Figure 2.1 indicate strains which are approximately proportional to stress at low stress levels, suggesting a large component of elastic distortion. If the stress is reduced the unloading stress-strain curve indicates that not all the strain is recovered on unloading. The hysteresis loss represents the energy lost in crushing and repositioning of grains. At higher shear stresses the strains are proportionally greater indicating greater crushing and reorientation. Indeed Arnold and Mitchell[8] showed that as a sample of sand is subjected to cyclic loading, the unloading response involves an increasing degree of hysteresis; in other words they found that on unloading recoverable deformation in sand under triaxial conditions was small. Because irrecoverable strains were larger than the elastic strains this led them to suggest that total strains could, in fact, be regarded as irrecoverable. As would be expected loose sand with larger voids and fewer points of contact exhibits greater strains and less recovery when unloaded than dense sand (see Lambrechts and Leonards[9], Lade[10]).

The angle of shearing resistance is also influenced by the grain size distribution

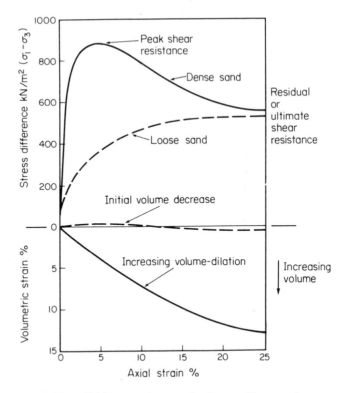

Figure 2.1 Stress-strain curves for dense and loose sand

and grain shape (see Holtz and Gibbs[11]); the larger the grains the wider the zone affected, the more angular the grains the greater the frictional resistance to their relative movement, since they interlock more thoroughly than do rounded ones and they therefore produce a larger angle of shearing resistance (Table 2.1). A well graded granular soil experiences less breakdown on loading than a uniformly sorted soil of the same mean particle size since in the former type there are more interparticle contacts and hence the load per contact is less than in the latter. In gravels the effect of angularity is less because of particle crushing.

Kirkpatrick[13]; Sutherland and Mesdary[14] and Arnold and Mitchell[8] all studied the failure state of sand in a three-dimensional stress system. They found that generally the Mohr-Coulomb law for sand based on triaxial tests under-predicts the failure strength of the material in other stress conditions.

Table 2.1 EFFECT OF GRAIN SHAPE AND GRADING ON THE PEAK FRICTION ANGLE OF COHESIONLESS SOIL (after Terzaghi[12])

Shape and grading	Loose	Dense
1. Rounded, uniform	30°	37°
2. Rounded, well graded	34°	40°
3. Angular, uniform	35°	43°
4. Angular, well graded	39°	45°

Table 2.2 SOME VALUES OF THE COMMON PROPERTIES OF SOILS

A. COHESIONLESS SOILS

	Gravels	*Sands*
Specific gravity	2.5–2.8	2.6–2.7
Bulk density (Mg/m^3)	1.45–2.3	1.4–2.15
Dry density (Mg/m^3)	1.4–2.1	1.35–1.9
Porosity (%)	20–50	23–35
Shear strength (kPa)	200–600	100–400
Angle of friction	35–45°	32–42°

B. COHESIVE SOILS

	Silts	*Clays*
Specific gravity	2.64–2.66	2.55–2.75
Bulk density (Mg/m^3)	1.82–2.15	1.5–2.15
Dry density (Mg/m^3)	1.45–1.95	1.2–1.75
Void ratio	0.35–0.85	0.42–0.96
Liquid limit (%)	24–35	Over 25
Plastic limit (%)	14–25	Over 20
Coefficient of consolidation (m^2/yr)	12.2	5–20
Effective cohesion (kPa)	75	20–200
Effective angle of friction	32–36°	

C. ORGANIC SOILS AND FILL

	Peat	*Coarse discard*
Moisture content (%)	650–1100	6–14
Specific gravity	1.3–1.7	1.8–2.7
Bulk density (Mg/m^3)	0.91–1.05	1.2–2.4
Dry density (Mg/m^3)	0.07–0.11	1.05–2.0
Void ratio	12.7–14.9	0.35–Over 1
Liquid limit (%)		23–45
Plastic limit (%)		Non-plastic–35
Effective angle of friction	5°	28°–40°
Effective cohesion (kPa)	20	20–50

The presence of water in the voids of a granular soil does not usually produce significant changes in the value of the angle of internal friction. However, if stresses develop in the pore water they may bring about changes in the effective stresses between the particles whereupon the shear strength and the stress-strain relationships may be radically altered. Whether or not pore pressures develop depends upon the drainage characteristics of the soil mass and its tendency to undergo volume changes when subjected to stress. If the pore water can readily drain from the soil mass during the application of stress then the granular material behaves as it does when dry. On the other hand if loading takes place rapidly, particularly in fine grained sands which do not drain as easily, then pore pressures are not dissipated. Since the water cannot readily escape from the voids of loosely packed, fine grained sands no volume decrease can occur and so the pressure increases in the pore water. If the sample is loose enough nearly all the stress difference may be carried by the pore water so that very little increase occurs in the effective stress. In dense sands, if the stress and drainage conditions prevent the water flowing into the sand as it is stressed then the usual volume increase characteristic of dense dry sand does not occur and a negative pore pressure develops.

The relationship between unit load on a potential surface of sliding (*p*) and the shearing resistance per unit area (s) can be approximately expressed by the equation:

$$s = (p - u) \tan \phi$$

where u is the hydrostatic pressure of the pore liquid prior to the application of the shearing force.

In the light of what has been said above concerning changing pore pressures this expression can be modified to:

$$s = (p - u - \Delta u) \tan \phi$$

where Δu represents the change in pore pressure.

Dusseault and Morgenstern[15] introduced the term 'locked sands' to describe certain peculiar sands which were first recognised in the Athabasca Oil Sands in Canada, and are older than Quaternary age. They are characterised by their high quartzose mineralogy, lack of interstitial cement, low porosity, brittle behaviour and high strength, with residual shear strengths (ϕ_r) varying between 30° and 35°.

Locked sands possess very high densities. Indeed the density indices of locked sands exceed 100%, that is, their porosities are less than those which can be obtained by laboratory tests for achieving minimum porosity. Dusseault and Morgenstern attributed this to the peculiar fabric of these sands. This has been developed by diagenetic processes which reduced the porosity of the sands by solution and recrystallisation of quartz as crystal overgrowths. Hence locked sands have an interlocked texture with a relatively high incidence of long and interpenetrative grain contacts. Sutured contacts are present and all grains have rugose surfaces.

At low stress levels locked sands undergo high rates of dilation. They have peak frictional strengths considerably in excess of those of dense sand. Dilatancy becomes suppressed as the level of stress increases, since the asperities on the surfaces of individual grains are sheared through rather than causing dilation. The failure envelopes of locked sands are steeply curved as a result of the changing energy level required for shearing asperities as the level of stress increases.

If undisturbed, locked sands are capable of supporting large loads with only small deformations.

References

1. Glossop, R. and Skempton, A.W., 'Particle size in silts and sands, *J. Inst. Civ. Engrs.*, **25**, Paper No. 5492, 81–105 (1945).
2. Kahn, J.S., 'The analysis and distribution of the properties of packing in sand size particles' *J. Geol.*, **64**, 385–395, 578–606 (1956).
3. Graton, L.C. and Frazer, H.J., 'Systematic packing of spheres with particular relation to porosity and permeability', *J. Geol.*, **43**, 785–909 (1935).
4. Cornforth, D.H., 'Some experiments on the influence of strain conditions on the strength of sand,' *Geotechnique*, **14**, 143–167 (1964).
5. Borowicka, H., 'Rearrangement of grains by shear tests with sand,' *Proc. 8th Int. Conf. Soil Mech. Found. Engng., Moscow*, **1**, 71–77 (1973).
6. Rowe, P.W., 'The relation between the shear strength of sands in triaxial compression, plane strain and direct shear', *Geotechnique*, **19**, 75–86 (1969).
7. Frossard, E., 'Effect of sand grain shape on the interparticle friction; indirect measurements by Rowe's stress dilatancy theory., *Geotechnique*, **29**, 341–350 (1979).
8. Arnold, M. and Mitchell, P.W., 'Sand deformation in three-dimensional stress state', *Proc. 8th Int. Conf. Soil Mech. Found. Engng., Moscow*, **1**, 11–18 (1973).

9. Lambrechts, J.R. and Leonards, G.A., 'Effects of stress history on deformation of sand', *Proc. A.S.C.E., J. Geotech. Engng. Div.*, 104, GT11, 1371–1389 (1978).

10. Lade, V.P., 'Prediction of undrained behaviour of sand,' *Proc. A.S.C.E., J. Geotech. Engng. Div.*, 104, GT6, 721–736 (1978).

11. Holtz, W.G. and Gibbs, H.J., 'Shear strength of pervious gravelly soils', *Proc. A.S.C.E.*, 82, Paper No. 867, (1956).

12. Terzaghi, K., 'The influence of geological factors in the engineering properties of sediments,' *Econ. Geol.*, 50th Ann. Vol., 557–618 (1955).

13. Kirkpatrick, W.M., 'The condition of failure of sands,' *Proc. 4th Int. Conf. Soil Mech. Found. Engng., London*, 1, 172–185 (1957).

14. Sutherland, H. and Mesdary, M., 'The influence of the intermediate principal stress on the strength of sand,' *Proc. 7th Int. Conf. Soil Mech. Found. Engng.*, 1, 391–399 (1969).

15. Dusseault, M.B. and Morgenstern, N.R., 'Locked sands,' *Q. Jl. Engng. Geol.*, 12, 117–132 (1979).

Chapter 3

Cohesive Soils

3.1 SILTS

Silts are clastic sediments, i.e. they are derived from pre-existing rock types chiefly by mechanical breakdown processes. They are mainly composed of fine quartz material. Silts may occur in residual soil horizons but in such instances they are usually not important. However, silts are often found in alluvial, lacustrine and marine deposits. As far as alluvial sediments are concerned silts are typically present in flood plain deposits, they may also occur on terraces which border such plains. These silts tend to interdigitate with deposits of sand and clay. Silts are also present with sands and clays in estuarine and deltaic sediments.

Lacustrine silts are often banded and may be associated with varved clays, which themselves contain a significant proportion of particles of silt size. Marine silts are also frequently banded and have high moisture contents. Wind blown silts are usually uniformly sorted. Grains of silt are often rounded with smooth outlines which influence their degree of packing. The latter, however, is more dependent on the grain size distribution within a silt deposit, uniformly sorted deposits not being able to achieve such close packing as those in which there is a range of grain size. This, in turn, influences the porosity and void ratio values as well as the bulk and dry densities (Table 2.2).

Dilatancy is characteristic of fine sands and silts. The environment is all important for the development of dilatancy since conditions must be such that expansion can take place. What is more it has been suggested that the soil particles must be well-wetted and it appears that certain electrolytes exercise a dispersing effect thereby aiding dilatancy. The moisture content at which a number of sands and silts from British formations become dilatant usually varies between 16 and 35 per cent. According to Boswell[1], dilatant systems are those in which the anomalous viscosity increases with increase of shear.

Schultze and Kotzias[2] showed that consolidation of silt was influenced by grain size, particularly the size of the clay fraction, porosity and natural moisture content. Primary consolidation accounted for 76% of the total consolidation exhibited by the Rhine silts tested by Schultze and Kotzias, secondary consolidation contributing the remainder. It was noted that unlike many American silts, which are often unstable when saturated, and undergo significant settlements when loaded, the Rhine silts in such a condition were usually stable. The difference no doubt lies in the respective soil structures. Most American silts are, in

20

fact, loess soils which have a more open structure than the reworked river silts of the Rhine with a void ratio of less than 0.85. Nonetheless, in many silts, settlement continues to take place after construction has been completed and may exceed 100 mm. Settlement may continue for several months after completion because the rate at which water can drain from the voids under the influence of applied stress is slow.

Schultze and Horn[3] found that the direct shear test proved unsuitable for the determination of the shear strength of silt, this had to be obtained by triaxial testing. They demonstrated that the true cohesion of silt was a logarithmic function of the water content and that the latter and the effective normal stress determined the shear strength. The angle of friction was dependent upon the plasticity index.

In a series of triaxial tests carried out on silt Penman[4] showed that in drained tests with increasing strain, the volume of the sample first decreased, then increased at a uniform rate and ultimately reached a stage where there was no further change. The magnitude of the dilatancy which occurred when the silt was sheared, and was responsible for these volume changes, increased with increasing density as it does in sands. The expansion was caused by the grains riding over each other during shearing. The strength of the silt was attributed mainly to the friction between the grains and the force required to cause dilatancy against the applied pressures.

These drained tests indicated that the angle of shearing resistance (ϕ) decreased with increasing void ratio and with increasing lateral pressure. Grain interlocking was responsible for the principal increase in the angle of shearing resistance and increased with increasing density. A fall in pore water pressure occurred in the undrained tests during shearing, and there was an approximately linear relation between the maximum fall in pore pressure and the void ratio. Provided the applied pressures were sufficiently high the drop in pore pressure governed the ultimate strength. The fall in the pore pressure was dependent on the density of the silt, the greater the density, the greater the fall in pore pressure. At a given density the ultimate strength was independent of applied pressure (above a critical pressure) and so the silt behaved as a cohesive material ($\phi = 0$). Below this critical pressure silt behaved as a cohesive and frictional material. An exceptional condition occurred when a highly dilatable sample was placed under low cell pressure. When the pore pressure fell below atmospheric pressure gas was liberated by the pore water and the sample expanded.

Frost heave (see Chapter 9) is commonly associated with silty soils and loosely packed silts can exhibit quick conditions.

3.2 LOESS

Loess is a wind-blown deposit which is mainly of silt size and consists mostly of quartz particles with lesser amounts of feldspar and clay minerals. It is characterised by a lack of stratification and uniform sorting and occurs as blanket deposits in central and western Europe, USA, Russia and China. Loess deposits are of Pleistocene age and because they show a close resemblance to fine grained glacial debris their origin has customarily been assigned a glacial association. In other words winds blowing from the arid interiors of the northern continents during glacial times picked up fine glacial outwash material and carried it many hundreds of kilometres before deposition took place. Deposition is presumed to have occurred over steppe lands, the grasses having left behind fossil root-holes which typify loess. The lengthy transport accounts for the uniform sorting of loess.

As can be inferred from the previous paragraph, loess owes its engineering characteristics largely to the way in which it was deposited since this gave it a metastable structure, in that initially the particles were loosely packed. The porosity of the structure is enhanced by the presence of fossil root-holes, but the latter have been subsequently lined with carbonate cement, which also helps bind the grains together. This has meant that the initial metastable structure has been preserved and the carbonate cement provides the bonding strength of loess. It must be pointed out, however, that the chief binder is usually the clay matrix. In a detailed examination of the micro-structure of loess soils Larionov[5] found that the coarser grains were never in contact with each other, being carried in a fine granular dispersed mass. Hence the strength of the soil is largely determined by the character of this fine mass. The ratio of coarser grains to fine dispersed fraction varies not only quantitatively but morphologically. Consequently three micro-structures can be recognised, namely, granular, where a filmy distribution of the fine dispersed fraction predominates; aggregate, consisting mainly of aggregates; and granular-aggregate, having an intermediate character. Larionov suggested that generally loess soils with granular micro-structure have less water resistance than aggregate types, they also have lower cohesion and higher permeability. They are therefore more likely to collapse on wetting than the aggregate types.

Loess deposits generally possess a uniform texture consisting of 50–90% particles of silt size. Their liquid limit averages about 30. Liquid limits as high as 45 have been recorded, and the plasticity index ranges from about 4 to 9, but averages 6. As far as their angle of shearing resistance is concerned this usually varies from 30–34°. Loess deposits are better drained (their permeability ranges from 10^{-5} to 10^{-7} m/s than true silts because of their pattern of vertical root-holes. As would be expected the coefficient of permeability is appreciably higher in the vertical than horizontal direction.

In the unweathered state above the water table the unconfined compressive strength of loess may amount to several hundred kilonewtons a square metre. On the other hand, if loess is permanently submerged the metastable structure breaks down so that loess then becomes a slurry.

As noted above the metastable structure of loess is liable to collapse if wetted. As a consequence several collapse criteria have been proposed which depend upon the natural void ratio (e) and the void ratios at the liquid limit (e_1) and the plastic limit (e_p). According to Audric and Bouquier[6] collapse is probable when the natural void ratio is higher than a critical void ratio (e_c) which depends on e_1 and e_p. They quoted the Denisov and Feda criteria as providing fairly good estimates of the likelihood of collapse:

$$e_c = e_1 \text{ (Denisov)}$$

$$e_c = 0.85e_1 + 0.15e_p \text{ (Feda)}$$

Audric and Bouquier went on to describe a series of consolidated undrained triaxial tests they had carried out, at natural moisture content and after wetting, on loess soil from Roumare in Normandy. They distinguished collapsible and non-collapsible types of loess. The main feature of the collapsible soils was the soil structure, which was formed by soil domains with a low number of grain contacts. The authors noted that when collapsible loess was tested, the deviator stress reached a peak at rather small values of axial strain and then decreased with further strain. The pore pressures continued to increase after the peak deviator stress had been reached. By contrast, in non-collapsible soils the deviator stress continued to increase and there was only a small increase of pore pressure. As expected the shear strength of collapsible loess was always less than that of the non-collapsible type.

The presence of vertical root-holes explains why vertical slopes are characteristic of loess landscapes. These may remain stable for long periods and when failure occurs it generally does so in the form of a vertical slice. By contrast an inclined slope is subject to rapid erosion.

Unlike silt, loess does not appear to be frost susceptible, this being due to its more permeable character. However, like silt, it can exhibit quick conditions and it is difficult, if not impossible, to compact. Because of its porous structure a 'shrinkage' factor must be taken into account when estimating earthwork. A full account of loess as a foundation material is provided by Clevenger[7].

3.3 CLAY MINERALS

Clay deposits are principally composed of fine quartz and clay minerals. The latter represent the commonest breakdown products of most of the chief rock forming silicate minerals. Clay minerals are phyllosilicates and their atomic structure, which significantly affects their engineering behaviour, can be regarded as consisting of two fundamental units. One of these units is composed of

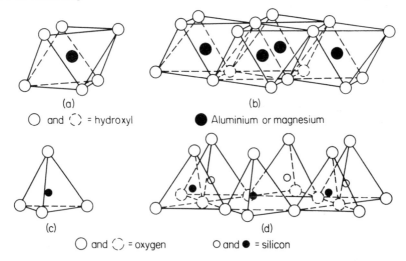

(a)

○ and ◌ = hydroxyl

(b)

● Aluminium or magnesium

(c)

○ and ◌ = oxygen

(d)

○ and ● = silicon

Figure 3.1 Fundamental units comprising the structure of clay minerals. (a) A single octahedral unit, in gibbsite Al is surrounded by 6 oxygens whereas in brucite they surround Mg in six-fold coordination; (b) the sheet structure of the octahedral units; (c) the silica tetrahedron; (d) the sheet structure of silica tetrahedrons arranged in a hexagonal network (After Grim[14])

two sheets of closely packed oxygens or hydroxyls in which atoms of aluminium, magnesium or iron are arranged in octohedral coordination (Figure 3.1). The other unit is formed of linked SiO_4 tetrahedrons which are arranged in layers (Figure 2.1b). In the common clay minerals these fundamental units are arranged in the respective atomic lattices shown in Figures 3.2a, b and c. The chemical

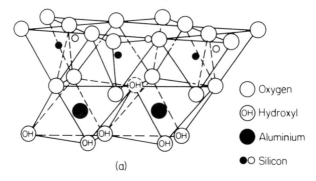

(a)

Figure 3.2a Diagrammatic sketch of the kaolinite structure

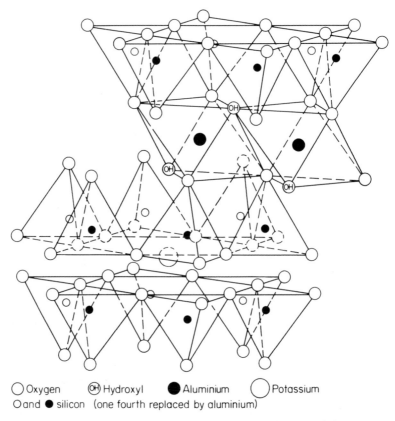

Figure 3.2b The above structure is that of muscovite which is regarded as essentially the same as that of illite

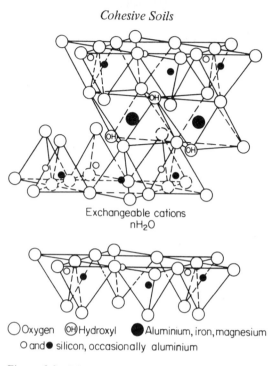

Exchangeable cations
nH₂O

◯ Oxygen ⊙Hydroxyl ⬤ Aluminium, iron, magnesium
O and ● silicon, occasionally aluminium

Figure 3.2c Diagrammatic sketch of the montmorillonite structure

composition of the clay minerals varies according to the amount of aluminium which is substituted for silicon in the atomic structure and also with the replacement of magnesium by other ions. Nevertheless they are basically hydrated aluminium silicates.

The three major clay minerals are kaolinite, illite and montmorillonite. Kaolinite is principally formed as an alteration product of feldspars, feldspathoids and muscovite as a result of weathering under acidic conditions. It is the most important clay mineral in china clays, ball clays and fireclays as well as most residual and transported clay deposits. Illite is a common mineral in most clays and shales, and is present in various amounts in tills and loess, but less common in soils. It develops due to the weathering of feldspars, micas and ferromagnesium silicates, or may form from other clay minerals upon diagenesis. Irrespective of the process responsible for its formation, this appears to be favoured by an alkaline environment.

Both kaolinite and illite have non-expansive lattices whilst that of montmorillonite is expansive. In other words montmorillonite is characterised by its ability to swell and by its notable cation exchange properties. The basic reason why montmorillonite can readily absorb water into the interlayer spaces in its sheet structure is simply that the bonding between them is very weak. Montmorillonite forms when basic igneous rocks, in badly drained areas, are weathered. An alkaline environment also favours its formation.

The shape, size and specific surface all influence the engineering behaviour of clay minerals. As noted above, clay minerals have a plate-like shape. They are very small in size, being measured in angstrom units. For example, an individual particle of montmorillonite is typically 1000 Å by 10 Å thick, whilst kaolinite is 10000 Å by 1000 Å thick. The specific surface refers to the magnitude of the surface in relation to the mass and the smaller the particle, the larger the specific surface (Table 3.1). The specific surface provides a good indication of the relative influence of electrical forces on the behaviour of a particle.

The surface of a clay particle has a nett charge. This means that a clay particle is surrounded by a strongly attracted layer of water, but as the diapolar water molecules do not satisfy the electrostatic balance at the surface of the clay particle, some metal cations are also adsorbed. The ions are usually weakly held and therefore can be readily replaced by others. Consequently they are referred to as exchangeable ions. The ion exchange capacity of soils normally ranges up to 40 milli-equivalents per 100 g. However, for some clay soils it may be greater as can be inferred from the ion exchange capacity of kaolinite, illite and montmorillonite (Table 3.1).

Table 3.1 SIZE AND SPECIFIC SURFACE OF SOIL PARTICLES

Soil particle	Size (mm)	Specific surface (m^2/g)	Ion exchange/capacity (Milli-equivalents/100 g)
Sand grain	1	0.002	
Kaolinite	$d = 0.3$ to 3^{-3} thickness = 0.3 to 0.1d	10–20	3–15
Illite	$d = 0.1$ to 2^{-3} thickness = 0.1d	80–100	20–40
Montmorillonite	$d = 0.1$ to 1^{-3} thickness = 0.01d	800	60–100

The type of adsorbed cations influences the behaviour of the soil in that the greater their valency, the better the mechanical properties. For instance, clay soils containing montmorillonite with sodium cations are characterised by high water absorption and considerable swelling. If these are replaced by calcium, a cation with a higher valency, both these properties are appreciably reduced. The thickness of the adsorbed layer influences the soil permeability, that is, the thicker the layer, the lower the permeability since a greater proportion of the pore space is occupied by strongly held adsorbed water. As the ion exchange capacity of a cohesive soil increases so does its plasticity index, the relationship between the two being almost linear.

3.4 MICRO-STRUCTURES

The micro-structure of cohesive soils is largely governed by the clay minerals present and the forces acting between them. Because of the complex electrochemistry of clay minerals the spatial arrangement of newly sedimented particles is very much influenced by the composition of the water in which deposition takes place. Single clay mineral platelets may associate in an edge-to-edge (EE), edge-to-face (EF) face-to-face (FF) or random type of arrangement depending on the interparticle balance between the forces of attraction and repulsion, and the amount or absence of turbulence in the water in which deposition occurs. Since sea water represents electrolyte rich conditions it causes clay particles to flocculate (see Lambe[8]). In other words the particles are attracted to one another in a loose, haphazard arrangement and considerable free water is trapped in the large voids (Figure 3.3a). Lambe[9] suggested that the clay particles were typically arranged in edge-to-edge or edge-to-face associations.

Flocculent soils are light in weight and very compressible but are relatively strong and insensitive to vibration because the particles are lightly bound by their edge-to-face attraction. They are sensitive to remoulding which destroys the bond between the particles so that the free water is released to add to the adsorbed layers at the former points of contact. By contrast, flocculation does not occur amongst clay particles deposited in fresh water. In this case they assume a more or less parallel, close-packed type of orientation. This has been referred to as a dispersed micro-structure (Figure 3.3b). The bulky grains are distributed throughout the mass and cause localised departures from the pattern. Soils having a dispersed structure are usually dense and watertight. Typical void ratios are often as low as 0.5. Estuarine clays, because they have been deposited in marine through brackish to freshwater conditions can contain a mixture of flocculated and dispersed micro-structures. Van Olphen[10] showed that the micro-structure of clay deposits was somewhat more complex than that proposed by Lambe[8]. The aggregate micro-structure (Figure 3.3c) was also proposed by Van Olphen[10].

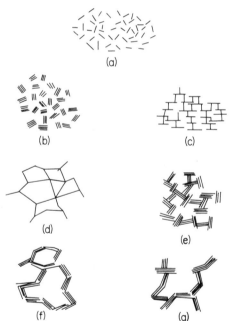

Figure 3.3 Modes of particle association in clay suspensions (After Van Olphen[10]) (a) Dispersed and flocculated (b) Aggregated but deflocculated (c) Edge-to-face flocculated but dispersed (d) Edge-to-edge flocculated but dispersed (e) Edge-to-face flocculated and aggregated (f) Edge-to-edge flocculated and aggregated (g) Edge-to-face and edge-to-edge flocculated and aggregated

The original micro-structure of a clay deposit is subsequently modified by overburden pressures due to burial, which bring about consolidation. Consolidation tends to produce a preferred orientation with the degree of reorientation of the clay particles being related to both the intensity of stress and the electro-chemical environment, dispersion encouraging and flocculation discouraging clay particle parallelism. For instance, Barden[11] maintained that lightly consolidated marine clay (e.g. Romerike Clay) retains a random open structure, that medium consolidated brackish water clay (e.g. Boston Blue Clay) has a very high degree of orientation, and that extremely heavily consolidated marine clay (e.g. London Clay) develops a fair degree of orientation.

It is generally accepted that turbostratic groups or domains occur in consolidated clays although it has been suggested that ill-defined domains are present in unconsolidated clays (see Aylmore and Quirk[12]; Burnham[13]). These domains consist of aggregations of clay particles which have a preferred orientation but between the aggregates the orientation is random. With increasing overburden pressure it appears that the number of clay particles in each domain increases and that there is an increase in domain orientation. For example, it has been shown that most of the Oxford Clay is made up of small domains, up to 0.005 mm across, these showing some alignment with the bedding.

3.5 PERFORMANCE OF CLAY DEPOSITS

The engineering performance of clay deposits is very much affected by the total water content and by the energy with which this moisture is held. For example, the moisture content influences their density, consistency and strength, and the energy with which moisture is held influences their volume change characteristics since swelling, shrinkage and consolidation are affected by permeability and moisture migration. Furthermore moisture migration may give rise to differential movement in clay soils. The gradients which generate moisture migration in clays may arise from variations in temperature, extent of saturation, and chemical composition or concentration of pore solutions. In order to minimise the deleterious effects of moisture movements in cohesive soils footings can be placed at depths which are unaffected by seasonal fluctuations of moisture content.

The capillary potential or soil water potential is the force required to pull a unit mass of water away from a unit mass of soil, its magnitude indicating the force with which moisture is held. Moisture moves from wet to dry clay and will move upwards under the influence of capillary action against the force of gravity until equilibrium is established. At equilibrium the pore water pressures decrease linearly with height above the water table.

Fully saturated clay soils often behave as incompressible materials when subjected to rapid loading. The amount of elasticity increases continuously as the water content is decreased. Elastic recovery of the original size or shape may be immediate or may take place slowly. The linear relationship between stress and strain only applies to clays at low stresses.

3.6 PLASTICITY

There is no particular value of plastic limit that is characteristic of an individual clay mineral, indeed the range of values for montmorillonite is large. This is due to the inherent variations of structure and composition within the crystal lattice and the variations in exchangeable-cation composition. Generally the plastic limits for the three clay minerals noted decrease in the order montmorillonite, illite and kaolinite. As far as montmorillonite is concerned if the exchangeable ions are Na and Li then these give rise to high plastic limits. In the case of the other two clay minerals the exchangeable cations produce relatively insignificant variation in the plastic limit. On the other hand poorly crystalline kaolinite of small particle size has a substantially higher plasticity than that of relatively coarse, well-organised particles.

Similarly there is no single liquid limit which is characteristic of a particular clay mineral; indeed the range of limits is much greater than that of the plastic limits. Again the highest liquid limits are obtained with Li and Na montmorillonite; then follow, in decreasing order, Ca, Mg, K, Al montmorillonite; illite; poorly crystalline kaolinite; well crystallised kaolinite. Indeed the liquid limit of Li and Na montmorillonite cannot be determined accurately because of their high degree of thixotropy. The character of the cation is not the sole factor influencing the liquid limit, the structure and composition of the silicate lattice are also important. The liquid limits for illites fall in the range 60 to 90 whilst those for kaolinites vary from about 30 to 75. Again the crystallinity of the lattice and particle size are the controlling factors, for instance, poorly crystallised, fine grained samples may be over 100. The presence of 10% montmorillonite in an illitic or kaolinitic clay can cause a substantial increase in their liquid limits.

The plasticity indices of Na and Li montmorillonite clays have exceedingly high values, ranging between 300 and 600. Montmorillonites with other cations have values varying from about 50 to 300 with most of them in the range of 75 to 125. As far as the latter are concerned there is no systematic variation with cation composition. In the case of illitic clays the plasticity indices range from 25 to 50. The values for well crystallised illite are extremely low, indeed they are almost non-plastic. The presence of montmorillonite in these clays substantially increases the index. The range of plasticity indices for kaolinitic clays varies from about 1 to 40, generally being around 25. As can be inferred from above the limit values increase with a decrease in particle size and the liquid limit tends to increase somewhat more than the plastic limit.

3.7 SWELLING AND SHRINKAGE

One of the most notable characteristics of clays from the engineering point of view is their susceptibility to slow volume changes which can occur independently of loading due to swelling or shrinkage. The ability of a clay to imbibe water leads to it swelling, and when it dries out it shrinks. Such movements in heavy clays in south-east England have been responsible for appreciable damage to buildings (see Tomlinson[14]).

Differences in the periods and magnitude of precipitation and evaporation are the major factors influencing the swell-shrink response of an active clay beneath a structure. Poor surface drainage and leakage from underground pipes can produce concentrations of moisture in clay. Trees with high water demand and uninsulated hot-process foundations may dry out a clay causing shrinkage. The density of a clay soil also influences the amount of swelling it is likely to undergo. Expansive clay minerals absorb moisture into their lattice structure (see below), tending to expand into adjacent zones of looser soil before volume increase occurs. In a densely packed soil having small void space, the soil mass has to swell to accommodate the volume change of the expansive clay particles.

Grim[15] distinguished two modes of swelling in clay soils, namely, intercrystalline and intracrystalline swelling. Interparticle swelling takes place in any type of clay deposit irrespective of its mineralogical composition; the process is reversible. In relatively dry clays, the particles are held together by relict water under tension from capillary forces. On wetting, the capillary force is relaxed and the clay expands. In other words intercrystalline swelling takes place when the uptake of moisture is restricted to the external crystal surfaces and the void spaces between the crystals. Intracrystalline swelling, on the other hand, is characteristic of the smectite family of clay minerals, and of montmorillonite in particular. The individual molecular layers which make up a crystal of montmorillonite are weakly bonded so that on wetting, water enters not only between the crystals but also between these unit layers which comprise the crystal. Swelling in Na montmorillonite is the most notable and can amount up to 1000% of the original volume, the clay then having formed a gel.

Hydration volume changes are frequently assessed in terms of the free-swell capacity. Mielenz and King[16] showed that generally kaolinite has the smallest swelling capacity of the clay minerals and that nearly all of its swelling is of the interparticle type. Illite may swell by up to 15% but intermixed illite and montmorillonite may swell some 60–100%. Swelling in Ca montmorillonite is very much less than in the Na variety, it ranges from about 50–100%. The large swelling capacity of montmorillonite means that they give the most trouble in foundation work.

Cycles of wetting and drying are responsible for slaking in argillaceous sediments which can bring about an increase in their plasticity index and augment their ability to swell. The air pressure in the pore spaces helps the development of the swell potential under cyclic wetting and drying conditions. On wetting, the pore air pressure in a dry clay increases and it can become large enough to cause breakdown, which at times can be virtually explosive. The rate of wetting is important, slow wetting allowing the air to diffuse through the soil water so that the pressure does not become large enough to disrupt the soil. In weakly bonded clay soils, cyclic wetting and drying brings about a change in the swell potential as a result of the breakdown of the bonds between clay minerals and the alteration of the soil structure.

Bjerrum[17] suggested that clay soils with high salt contents and a network of cracks undergo an increase in swell potential due to slaking consequent upon osmotic pressures being developed as rain water infiltrates into the cracks.

Freeze-thaw action and osmotic swelling also affect the swell potential. Freezing can give rise to large internal pressures at the freezing front in fine grained soils and if this front advances slowly enough the soil immediately beneath it can become quite desiccated. On melting the desiccation zone is saturated. Hence the swell potential is increased by the freeze pressures.

According to Schmertmann[18] some clays increase their swell behaviour when they undergo repeated large shear strains due to mechanical remoulding. He introduced the term swell sensitivity for the ratio of the remoulded swelling index to the undisturbed swelling index and suggested that such a phenomenon may occur in unweathered highly overconsolidated clay when the bonds, of various origins, which hold clay particles in bent positions have not been broken.

When these bonds are broken by remoulding, the clays exhibit significant swell sensitivity.

An internal swelling pressure will reduce the effective stress in a clay soil and therefore will reduce its shearing strength. As a consequence Hardy[19] suggested that the Coulomb equation for swelling soils should take the swelling pressure as well as pore pressure into account, it becoming

$$\tau = c + (\sigma - u - p_s) \tan \phi'$$

where

τ is the shearing strength;

c is the cohesion;

σ is the total stress;

u is the pore water pressure;

p_s is the swelling pressure and

ϕ' is the angle of effective internal friction.

Holtz and Gibbs[20] showed that expansive clays can be recognised from their plasticity characteristics and Holtz[21] suggested a relationship between colloid content, plasticity index and shrinkage limit to indicate the degree of expansion of certain clay soils. However, the most widely used soil property to predict swell potential is the activity of a clay (see Van der Merwe[22]).

The plasticity of a cohesive soil is influenced by the amount of its clay fraction since clay minerals greatly influence the amount of attracted water held

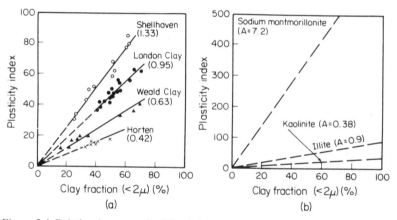

Figure 3.4 Relation between plasticity index and clay fraction. Figures in brackets represent the activities of the clays. (After Skempton[23])

in a soil. With this in mind Skempton[23] defined the activity of a clay deposit as the ratio of the plasticity index to the percentage of the clay fraction:

$$\text{Activity} = \frac{\text{Plasticity Index}}{\% \text{ by weight finer than 0.002 mm}}$$

Skempton suggested three classes of activity, namely, active, normal and inactive which he further subdivided into five groups as follows:

1. Inactive with activity less than 0.5
2. Inactive with activity range 0.5 to 0.75
3. Normal with activity range 0.75 to 1.25

4. Active with activity range 1.25 to 2
5. Active with activity greater than 2.

The activity of many British soils varies between 0.75 and 1.25, the exceptions being lacustrine and estuarine clays which tend to be lower and higher respectively. It would appear that there is only a general correlation between the clay mineral composition of a deposit and its activity, that is, kaolinitic and illitic clays are usually inactive whilst montmorillonitic clays range from inactive to active (Figure 3.4). Usually active clays have a relatively high water holding capacity and a high cation exchange capacity. They are also highly thixotropic, have a low permeability and a low resistance to shear.

Some engineering problems associated with expansive clays have been reviewed by Popescu[24].

3.8 HEAVE POTENTIAL

Volume changes in clays also occur as a result of loading and unloading which bring about consolidation and heave respectively. When clay is first deposited in water, its water content is very high and it may have void ratios exceeding 2. As sedimentation continues overburden pressure increases and according to Hedberg[25] the particles are rearranged to produce a new packing mode of greater stability as most of the water is expelled from the deposit. He suggested that the expulsion of free and adsorbed water took place until a porosity of about 30% was reached. Any further reduction in pore volume he attributed to mechanical deformation of particles or diagenetic processes.

As material is removed from a deposit by erosion the effective overburden stress is reduced and elastic rebound begins. Part of the rebound or heave results from an increase in the water content of the clay. Cyclic deposition and erosion has resulted in multiple loading and unloading of many clay deposits.

The heave potential arising from stress release depends upon the nature of the diagenetic bonds within the soil, that is, the post-depositional changes such as precipitation of cement and recrystallization which have occurred (see Bjerrum[17]). It would appear that significant time dependent vertical swelling may arise, at least in part, from either of two fundamentally different sources, namely, localised shear stress failures or localised tensile stress failures. Localised shear stress failures are associated with long term deformations of soils having weak diagenetic bonds. Localised tensile stress failures are probably associated with long term deformations of soils having well developed diagenetic bonds such as clay shales.

When an excavation is made in a clay with weak diagenetic bonds, elastic rebound will cause immediate dissipation of some stored strain energy in the soil. However, part of the strain energy will be retained due to the restriction on lateral straining in the plane parallel to the ground surface. The lateral effective stresses will either remain constant or decrease as a result of plastic deformation of the clay as time passes (see Bjerrum[26]). These plastic deformations can result in significant time dependent vertical heaving. However, creep of weakly bonded soils is not a common cause of heaving in excavations.

The relationships between the stresses, failure mechanisms and time dependent heaving are complex in clay soils with well developed diagenetic bonds. According to Obermeier[27] heaving is in part related to crack development, cracking giving rise to an increase in volume, the rate of crack growth being of particular significance. The initial rate of heaving is probably controlled by cracking due to tensile failures and to plastic deformations arising from shear failures. Furthermore Obermeier maintained that because of the breakdown of diagenetic bonds there is an increase in the lateral stresses parallel to the ground surface.

When a load is applied to a clay soil its volume is reduced, this being due principally to a reduction in the void ratio. If such a soil is saturated then the load is initially carried by the pore water which causes a pressure, termed the hydrostatic excess pressure, to develop. The excess pressure of the pore water is dissipated at a rate which depends upon the permeability of the soil mass and the load is eventually transferred to the soil structure. The change in volume during consolidation is equal to the volume of the pore water expelled and corresponds to the change in void ratio of the soil. In clay soils, because of their low permeability, the rate of consolidation is slow. Consolidation brought about by a reduction in the void ratio is termed primary consolidation. Further consolidation may occur due to a rearrangement of the soil particles. This is much less significant and is referred to as secondary consolidation. The various factors which influence the compressibility of a clay soil have been reviewed by Wahls[28].

El Refai and Hsu[29] have provided a recent discussion of creep deformation in clays.

3.9 SETTLEMENT AND BEARING CAPACITY OF CLAY SOILS

For all types of foundation structures on clays the factors of safety must be adequate against bearing capacity failures. Experience has indicated that it is desirable to use a factor of safety of 3, yet although this will eliminate complete failure, settlement may still be excessive. It is, therefore, necessary to give consideration to the settlement problem if bearing capacity is to be viewed correctly. More particularly it is important to make a reliable estimate of the amount of differential settlement that may be experienced by the structure. If estimated differential settlement is excessive it may be necessary to change the layout or type of foundation structure.

During the construction period the net settlement is comprised of immediate settlement due to deformation of the clay without a change in water content and consolidation settlement brought about by pore water being squeezed from the clay (see Chang, Broms and Peck[30]). As mentioned above the rate of consolidation is generally very slow because of the low permeability of clays so that the former type of settlement will have been the greater of the two by the end of the construction period. In the course of time consolidation becomes important, giving rise to long continued settlement, although at a decreasing rate for years or decades after the completion of construction. Accordingly the principal objects of a settlement analysis are, firstly, to obtain a reasonable estimate of the net final settlement corresponding to a time when consolidation is virtually complete and, secondly, to estimate the progress of settlement with time. It should be borne in mind that settlement depends primarily on the compressibility of the clay which is, in turn, intimately related to its geological history, that is, to whether it is normally consolidated or overconsolidated (see below).

Clays which have undergone volume increase due to swelling or heave are liable to suffer significantly increased gross settlement when they are subsequently built upon.

A normally consolidated clay is that which, at no time in its geological history, has been subject to pressures greater than its existing overburden pressure, whereas an overconsolidated clay has been so subjected. The major factor in overconsolidation is removal of material that once existed above a clay deposit by erosion. Berre and Bjerrum[31] carried out a series of triaxial and shear tests on normally consolidated clay, the confining conditions simulating the overburden pressures in the field. They demonstrated that the clay could sustain a shear stress in addition to the *in situ* value, undergoing relatively small deformation as long as the shear did not exceed a given critical value. This critical shear value represents the maximum shear stress which can be mobilised under undrained conditions, and governs the bearing pressure such a clay can carry with limited amount of settlement.

Generally this critical shear value varies with plasticity and the rate at which load is applied. An overconsolidated clay is considerably stronger at a given pressure than a normally consolidated one, and it tends to dilate during shear whereas a normally consolidated clay consolidates. In both normally consolidated and overconsolidated clays the shear strength reaches a peak value and then, as displacements increase, decreases to the residual strength. The development of residual strength is therefore a continuous process.

The ultimate bearing capacity of foundations on clay soil depends on the shear strength of the soil and the shape and depth at which the foundation structure is placed (see Skempton[32]). Although there is a small decrease in the moisture content of a clay beneath a foundation structure, which gives rise to a small increase in soil strength, this is of no importance as far as estimation of the factor of safety against shear is concerned. In relation to applied stress saturated clays behave as purely cohesive materials provided that no change of moisture content occurs. Thus when a load is applied to saturated clay it produces excess pore pressures which are not quickly dissipated. In other words the angle of shearing resistance (ϕ) is equal to zero. The assumption that $\phi = 0$ forms the basis of all normal calculations of ultimate bearing capacity in clays (see Skempton[32]). The strength may then be taken as the undrained shear strength or one half the unconfined compressive strength. To the extent that consolidation does occur, the results of analyses based on the premise that $\phi = 0$ are on the safe side. Only in special cases, with prolonged loading periods or with very silty clays, is the assumption sufficiently far from the truth to justify a more elaborate analysis.

Lambe's[33] concept of the shear strength in clay postulated the existence of forces of attraction and repulsion between the particles with a net repulsive force in accordance with physico-chemical principles. Hence the equilibrium of internal stresses in a clay soil can be expressed as:

$$\sigma' = \sigma - u = R - A$$

where
    σ' is the effective stress;
    σ is the total stress;

u is the pore water pressure;

R is the total force of repulsion and

A is the total force of attraction per unit area between the particles.

Unfortunately the R and A forces cannot be measured. The relationship is demonstrated by the fact that clay behaviour differs when immersed in solutions with different electrolyte concentrations. For a discussion of the shear strength of saturated clays see Sridharan and Rao[34].

3.10 SENSITIVITY OF CLAYS

The shear strength of an undisturbed clay is generally found to be greater than that obtained when it is remoulded and tested under the same conditions and at the same water content. The ratio of the undisturbed to the remoulded strength at the same moisture content was defined by Terzaghi[35] as the sensitivity of a clay. Subsequently Skempton and Northey[36] proposed the following grades of sensitivity:

1. Insensitive clays, under 1
2. Low sensitive clays, 1 to 2
3. Medium sensitive clays, 2 to 4
4. Sensitive clays, 4 to 8

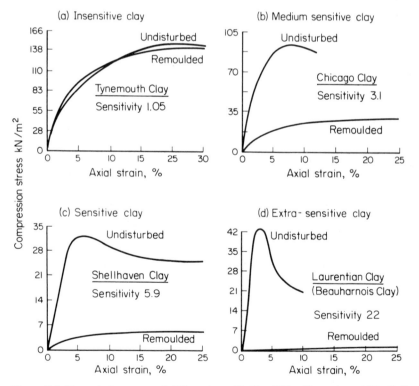

Figure 3.5 Stress-strain curves of different sensitivities (After Skempton, A.W., Soil mechanics in relation to geology, *Proc. Yorks. Geol. Soc.* 29 (1953))

5. Extra-sensitive clays, 8 to 16

6. Quick clay, over 16

Clays with high sensitivity values have little or no strength after being disturbed. Indeed if they suffer slight disturbance this may cause an initially fairly strong material to behave as a viscous fluid. High sensitivity seems to result from the metastable arrangement of equidimensional particles. The strength of the undisturbed clay is chiefly due to the strength of the framework developed by these particles and the bonds between their points of contact. If the framework is destroyed by remoulding, the clay loses most of its strength and any subsequent gain in strength due to thixotropic hardening does not exceed a small fraction of its original value. Sensitive clays generally possess high moisture contents, frequently with liquidity indices well in excess of unity. A sharp increase in moisture content may cause a great increase in sensitivity, sometimes with disastrous results. Heavily overconsolidated clays are insensitive. The effect of remoulding on clays of various sensitivities is illustrated in Figure 3.5.

Some clays with moderate to high sensitivity regain their strength when, after remoulding, they are allowed to rest under unaltered external conditions. Such soils are thixotropic, i.e. they undergo an isothermal gel-to-sol-to-gel transformation upon agitation and subsequent rest. This transformation can be repeated indefinitely without fatigue and the gelation time under similar conditions remains the same. The softening and subsequent recovery of thixotropic soils appears to be due, firstly to the destruction, and then secondly to the rehabilitation, of the molecular structure of the adsorbed layers of the clay particles. For example, the loss of consistency in soils containing Na montmorillonite occurs because large volumes of water are adsorbed upon and held between the colloidal clay particles.

Furthermore, the ionic forces attracting the colloidal clay particles together have a definite arrangement which is an easily destroyable microstructure when subjected to agitation. When the material is at rest, the ions and water molecules tend to reorientate themselves and strength is thereby recovered.

3.11 FISSURES IN CLAY

Many clays are weakened by a network of fissures. Terzaghi[37] provided the first quantitative data relating to the influence of fissures and joints on the strength of clays, pointing out that such features are characteristic of overconsolidated clays. He[38] maintained that fissures in normally consolidated clays have no significant practical consequences and that they appeared to be due to a process comparable with syneresis. On the other hand they can have a decisive influence on the engineering performance of an overconsolidated clay, in that the overall strength of such fissured clay can be as low as one tenth that of the intact clay.

Skempton[39] attributed the reduction in strength in the London Clay exposed in cuttings to softening along fissures which opened as a result of small movements consequent upon the removal of lateral support on excavation. In addition to allowing clay to soften, fissures and joints allow concentrations of shear stress, which locally exceed the peak strength of the clay, thereby giving rise to progressive failure. Under stress, the fissures in clay seem to propagate and coalesce in a complex manner. In most instances slope failures in cuttings in stiff fissured clay take place several years or decades after excavation but where fissures are closely spaced failure can occur during excavation. In fact it is not unusual for slopes in fissured clay to fail even though they were designed on strengths lower than the undrained strength parameters for intact clay. It appears that the average shearing resistance of stiff fissured clay at the moment of sliding usually ranges between 15 and 30 kPa whereas the initial shearing resistance of such clays ranges between 100 and 300 kPa. Slides in such clay may occur with little or no warning. Perhaps the best way to take care of this is to apply a higher factor of safety to undrained stability analyses. With emphasis on the probable area of fissures along a potential failure plane it should be possible to establish the total resistance to sliding offered by a fissured material by distinguishing those fissures likely to be open and offering no resistance from those closed and offering residual strength.

Skempton, Schuster and Petley[40] noted that the joints in London Clay were planar and that their orientation was predominantly normal to the horizontal bedding. They range up to 2.6 m in height and up to 6.0 m long. On the other hand the fissures may be curved or planar, or rarely more than 150 mm in size and exhibit no preferred orientation although they tend to be inclined at low angles to bedding planes. The mean size of the fissures decreases and the number per unit volume correspondingly increases as the upper surface of the clay is approached, suggesting that stress release and weathering play an important role in fissure formation. Similarly Ward, Marsland and Samuels[41] found that at Ashford the more weathered brown London Clay was more fissured than the blue clay below.

However, this concept has been disputed by some workers who have assumed that fissures are original features which are opened up on exposure. Fookes and Denness[42] suggested that the intensity of fissuring was influenced by near surface desiccation cracks and that the bedding planes appeared to be the major factor governing the fissure patterns which developed. They claimed that vertical stress release seemed to have little influence except when almost parallel to the bedding direction. But when this does happen they supposed that dilation occurs within the upper metre or so of the clay mass due to elastic expansion of the clay material and separation of individual lithological units. Fookes and Denness maintained that fissuring sympathetic to the bedding planes might have developed as a result of differential settlement between neighbouring particles shortly after deposition, leading to their separation with the formation of discontinuities approximately normal to the bedding. Minor changes in lithology do not appear to influence the attitude of fissuring but affect the size of fissures and the intensity of fissuring. Generally with increasing time after exposure, the proportion of non-planar to planar fissures increases since the former can develop from the latter by extension from their extremities.

Skempton, Schuster and Petley[40] summarised the shear strength parameters of the London Clay in terms of effective stress as follows:

Peak strength of intact clay

$$c' = 31 \text{ kPa}, \phi' = 20°$$

'Peak' strength on fissure and joint surfaces

$$c' = 6.9 \text{ kPa}, \phi' = 18.5°$$

Residual strength

$$c' = 1.4 \text{ kPa}, \ \phi'_r = 16°$$

Thus the strength along joints or fissures in clay is only slightly higher than its residual strength.

Similar results previously had been obtained by Marsland and Butler[43] from the stiff fissured Barton Clay. Hence the upper limit of the strength of fissured clay is represented by its intact strength whilst the lower limit is the strength along the fissures. The operational strength which is somewhere between the two is, however, often significantly higher than the fissure strength (see Lo[44]).

References

1. Boswell, P.G.H., *Muddy Sediments,* Heffer, Cambridge (1961).
2. Schultze, E. and Kotzias, A.B., 'Geotechnical properties of Lower Rhine Silts,' *Proc. 5th Int. Conf. Soil Mech. Found. Engng.,* 1, 329–333 (1961).
3. Schultze, E. and Horn, A., 'The shear strength of silt,' *Proc. 5th Int. Conf. Soil Mech. Found Engng,* 1, 350–358 (1961).
4. Penman, A.D.M., 'Shear characteristics of a saturated silt in triaxial compression,' *Geotechnique,* 3, 312–315 (1953).
5. Larionov, A.K., 'Structural characteristics of loess soils for evaluating their constructional properties,' *Proc. 6th Int. Conf. Soil Mech. Found. Engng., Montreal,* 64–68 (1965).
6. Audric, T. and Bouquier, L., 'Collapsing behaviour of some loess soils from Normandy,' *Q. J. Engng. Geol.,* 9, 265–278 (1976).
7. Clevenger, M.A., 'Experiences with loess as a foundation material,' *Trans. A.S.C.E.,* 123, 151–180 (Paper No. 2961), (1958).
8. Lambe, T.W., 'The structure of inorganic soil,' *Proc. A.S.C.E.,* 79, Separate No. 315 (1953).
9. Lambe, T.W., 'Structure of compacted clays,' *Proc. A.S.C.E. Soil Mech. Found. Engng. Div.,* SM4, 85–106 (1958).
10. Van Olphen, H., *An Introduction to Clay Colloid Chemistry,* Wiley, New York, (1963).
11. Barden, L., 'The relation of soil structure to the engineering geology of clay soil,' *Q.J. Engng. Geol.,* 5, 85–102 (1972).
12. Aylmore, L.A.G. and Quirk, J.P., 'Domain of turbostratic structures of clays,' *Nature,* 187, 1046–1048 (1960).
13. Burnham, C.P., 'Micromorphology of argillaceous sediments: particularly calcareous clays and siltstones,' article in *Micromorphological Techniques and Applications,* Ed. by Osmond, D.A. and Bullock, P. *Soil Surv. Tech. Mono.,* 2, 83–96 (1970).
14. Tomlinson, M.J., 'The design of foundation structures for difficult ground. In *Foundation Engineering in Difficult Ground,* ed. F.G.Bell, Newnes-Butterworths, London, 539–551 (1978).
15. Grim, R.E., *Applied Clay Mineralogy,* McGraw-Hill, New York (1962).
16. Mielenz, R.C. and King, M.E., 'Physical-chemical properties and engineering performance of clays.' In *Clays and Clay Technology,* ed. Pask, J.A. and Turner, M.D. California Division of Mines, Bulletin 169, 196–254 (1955).
17. Bjerrum, L., 'Progressive failure in slopes of overconsolidated plastic clay and clay shales,' *Proc. A.S.C.E. Soil Mech. Found. Engng. Div.,* SM5, 93, 2–49 (1967).
18. Schmertmann, J.H., 'Swell sensitivity,' *Geotechnique,* 19, 530–533 (1969).
19. Hardy, R.M., 'Identification and performance of swelling soil types,' *Canadian Geot. J.,* 2, 141–153 (1966).
20. Holtz, W.G. and Gibbs, H.J., 'Engineering properties of expansive clays,' *Trans. A.S.C.E.,* 121, 641–663 (1956).
21. Holtz, W.G., 'Expansive clays – properties and problems (In *Theoretical and practical treatment of expansive soils. Colorado School of Mines Quarterly,* 54, 89–117 (1959).

22. Van der Merwe, D.H., 'The prediction of heave from the plasticity index and the percentage clay fraction of soils. The Civil Engineer,' *S. Af. Inst. Civ. Engrs.*, **6**, 103–131 (1964).
23. Skempton, A.W., 'The colloidal activity of clays,' *Proc. 3rd Int. Conf. Soil Mech. Found. Engng., Zurich*, **1**, 57–61 (1953).
24. Peoescu, M.E., 'Engineering problems associated with expansive clays from Romania,' *Engng. Geol.*, **14**, 43–53 (1979).
25. Hedburg, H.D., 'Gravitational compaction of clays and shales,' *Am. J. Sci. (5th series)*, **31**, 241–287 (1936).
26. Bjerrum, L., 'Embankments on soft ground (In *Performance of earth and earth supported structures. Vol. II. A.S.C.E. Proc. Specialty Conf.*, Purdue Univ., Lafayette, Indiana, 32–33 (1972).
27. Obermeier, S.F., 'Evaluation of laboratory techniques for measurement of swell potential of clays.' *Bull. Ass. Engng. Geologists*, **11**, 293–314 (1974).
28. Wahls, H.E., 'Analysis of primary and secondary consolidation,' *Proc. ASCE J. Soil Mech. Found. Engng. Div.*, **88**, SM6, 207–231 (1962).
29. El Refai, W.T.H. and Hsu, J.R., 'Creep deformation of clays,' *Proc. A.S.C.E., J. Geotech. Engng. Div.*, **104**, GT, 61–76 (1978).
30. Chang, Y.C.E., Bro,ns, B. and Peck, R.B., 'Relationship between the settlement of soft clays and excess pore pressure due to improved loads,' *Proc. 8th Int. Conf. Soil Mech. Found. Engng., Moscow.* **1**, 93–86 (1973).
31. Berre, T. and Bjerrum, L., 'The shear strength of normally consolidated clays,' *Proc. 8th Int. Conf. Soil Mech. Found. Engng., Moscow*, **1**, 39–49 (1973).
32. Skempton, A.W., 'The bearing capacity of clays,' *Building Research Congress, Div. 1*, 180–189 (1951).
33. Lambe, T.W., 'A mechanistic picture of shear strength in clay,' *Proc. ASCE Res. Conf. Shear Strength Cohesive Soils*, Boulder, Colorado, 555–580 (1960).
34. Sridharan, A. and Venkatappa Rao, G., 'Shear strength behaviour of saturated clays and the role of the effective stress concept,' *Geotechnique*, **29**, 177–93 (1979).
35. Terzaghi, K., 'Ends and means in soil mechanics,' *Engineering J., (Canada)*, **27**, 608–615 (1944).
36. Skempton, A.W. and Northey, R.D., 'The sensitivity of clays,' *Geotechnique*, **2**, 30–53 (1952).
37. Terzaghi, K., 'Stability of slopes of natural clay,' *Proc. 1st Int. Conf. Soil Mech. Found. Engng.*, Cambridge, Mass., **1**, 161–165 (1936).
38. Terzaghi, K., 'The influence of modern soil studies on the design and construction of foundations,' *Building Research Congress. Div. 1*, 139–145 (1951).
39. Skempton, A.W., 'The rate of softening in stiff fissured clay with special reference to the London Clay,' *Proc. 2nd Ind. Conf. Soil Mech. Found. Engng.*, Rotterdam, **2**, 50–53 (1948).
40. Skempton, A.W., Schuster, R.L. and Petley, D.J., 'Joints and fractures in the London Clay at Wraysbury and Edgware,' *Geotechnique*, **19**, 205–217 (1969).
41. Ward, W.H., Marsland, A. and Samuels, S.G., 'Properties of the London Clay at the Ashford Common Shaft: *in situ* and undrained strength tests,' *Geotechnique*, **15**, 321–344 (1965).
42. Fookes, P.G. and Denness, B. 'Observational studies on fissure patterns in cretaceous sediments of south east England,' *Geotechnique*, **19**, 453–477 (1959).
43. Marsland, A. and Butler, M.E., 'Strength measurements on stiff fissured Barton Clay from Fawley, Hampshire,' *Proc. Geot. Conf. Oslo*, **1**, 139–146 (1967).
44. Lo, K.Y., 'The operational strength of fissured clay,' *Geotechnique*, **20**, 57–74 (1970).

Chapter 4

Soils Formed in Extreme Climates

4.1 CHARACTER AND TYPES OF TILLS

Till is regarded as being synonymous with boulder clay. It is deposited directly by ice whilst stratified drift or tillite is deposited in melt waters associated with glaciers. An extensive review of the various types of glacial deposits and their engineering properties has been provided by Fookes *et al.*[7].

The character of a till deposit depends on the lithology of the material from which it was derived, on the position in which it was transported in the glacier, and on the mode of deposition (see Boulton[2]; and Boulton and Paul[3]). The underlying bedrock usually constitutes up to about 80% of basal tills, depending on its resistance to abrasion and plucking. Argillaceous rocks such as shales, and mudstones are more easily abraded and produce fine grained tills which are presumably richer in clay minerals and therefore more plastic than other tills. Mineral composition also influences the natural moisture content which is slightly higher in tills containing appreciable quantities of clay minerals or mica. Upper tills have a high proportion of far-travelled material and may not contain any of the local bedrock.

Deposits of till consist of a variable assortment of rock debris ranging from fine rock flour to boulders. On the one hand they may consist essentially of sand and gravel with very little binder, alternatively they may have an excess of clay. Lenses and pockets of sand, gravel and highly plastic slickensided clay are frequently encountered in some tills. Most tills contain a significant amount of quartz in their silt-clay fractions. Frequently the larger, elongated fragments in till possess a general orientation in the path of ice movement.

The shape of the rock fragments found in till varies but is largely conditioned by the initial shape of the fragment at the moment of incorporation into the ice. Angular boulders are common, their irregular sharp edges resulting from crushing. Crushing or grinding of a rock fragment occurs when it comes in contact with another fragment or the rock floor.

Distinction has been made between tills derived from rock debris which was carried along at the base of a glacier and those deposits which were transported within and on the ice. The former is referred to as lodgement till whereas the latter is known as ablation till. Lodgement till is thought to be plastered onto the ground beneath the moving glacier in small increments as the basal ice melts. Because of the overlying weight of ice such deposits are overconsolidated.

Ablation till accumulates on the surface of the ice when englacial debris melts out, and as the glacier decays the ablation till is slowly lowered to the ground. It is therefore normally consolidated. Lodgement till contains fewer, smaller stones (they generally possess a preferred orientation) than ablation till and they are rounded and striated. Due to abrasion and grinding, the proportion of silt and clay size material is relatively high in lodgement till (e.g. the clay fraction varies from 15 to 40%). Lodgement till is commonly very compact and fissile, and is practically impermeable. It oxidises very slowly so that it is usually grey. Because it has not been subjected to much abrasion, ablation till is characterised by abundant large stones that are angular and not striated, the proportion of sand and gravel is high and clay is present only in small amounts (usually less than 10%). Because the texture is loose, ablation till oxidises rapidly and commonly is brown or yellowish brown. Since ablation till consists of the load carried at the time of ablation it usually forms a thinner deposit than lodgement till. Accordingly till sheets may comprise one or more layers of different materials, not all of which are likely to be found at any one locality. Shrinking and reconstitution of an ice sheet can complicate the sequence. In fact Elson[4] went further and divided tills into four categories, namely, superglacial ablation till, subglacial ablation till, comminution till and deformation till. Only superglacial ablation till corresponds with the definition of ablation till given above. It generally consists of thin lenses of sand and gravel of irregular distribution.

Englacial debris occurs mainly in the lower 30 to 60 m of a glacier where rock detritus may comprise as much as 10 or 20% of its volume. Consequently an appreciable thickness of englacial drift can be melted out from the base of a glacier although this suffers a reduction in volume of anything up to 90%. This subglacial ablation till may be precompressed by the overlying ice and may be sliced by thrust planes, dipping upstream near the glacier margin. Elongate stones may possess preferred orientations. The average grain size is much smaller than that of superglacial till. At the base of subglacial till deposits striated boulder pavements or thin irregular lenses of sand and pebbles may occur as relicts of the last erosive movement of the ice.

Comminution till occurs beneath subglacial ablation till and is formed by the shearing action of the ice at the base of the glacier. Elson suggested that this action generates enough heat to produce a sufficient quantity of melt water to bring about compaction to maximum density. Particles are oriented in the position of least resistance by the shearing action of moving ice and pebbles show a preferred orientation. Silt-sized tabular particles lie approximately horizontal or parallel to surfaces of larger particles, giving rise to microfoliation. Stones are surrounded by a compact matrix containing a high concentration of fines formed by abrasion of their surfaces.

Deformation or soft till may form from sandstone, siltstone, soft shale or weathered mantle. These materials cannot form dense comminution tills because the surfaces of the particles are not created by the glacier. These materials are porous and some initially contain more water than required for compaction to maximum density. Hence they yield soft tills which tend to be deformed rather than crushed.

McGown and Derbyshire[5] devised a more elaborate system for the classification of tills. Their classification is based upon the mode of formation, transportation and deposition of glacial material and provides a general basis for the prediction of the engineering behaviour of tills (Table 4.7; Figure 4.7). However,

Table 4.1 TILL TYPES AND PROCESSES (From McGown and Derbyshire[5])

Formative processes	Transportation processes	Depositional processes
Comminution till: Produced by abrasion and interaction between particles in the basal zone of a glacier. It is a common element in most tills.	Supraglacial till: Derived from frost riving of adjacent rocks or by differential melting out of glacier dirt beds. It may or may not become incorporated in the glacier and may suffer frost shattering and washing by melt-waters as it is transported on the top of the glacier.	*Ablation till: Accumulated by melting out on the surface of a glacier or as a coating on inert ice.
Deformation till: Produced by plucking, thrusting, folding and brecciation of the glacier bed.		*Meltout till: Accumulated as the ice of an ice-debris mixture melts out. Meltout tills exhibit a relatively low bulk density but show some variation depending on the particle size distribution. Generally the material is poorly sorted although occasionally thin layers or lenses of washed sediment occur. The microfabric is normally rather open. The clast fabrics of meltout tills show a wide variation depending on the disposition and debris concentration of the melting out mass.
	Englacial till: Derived from supraglacial till subsequently buried by accumulating snow or entrained in shear zones. It is transported within the ice mass and is more abundant in Polar regions than in temperate zones.	
	Basal till: Derived from comminution products in the ice-rock contact zones particularly the lowermost regions of a glacier. It is generally transported in concentrated bands in the bottom metre or so of a glacier.	Lodgement till: Accumulated subglacially by accretion from debris-rich basal ice. Lodgement tills generally possess a wide range of particle sizes and are frequently anisotropic, fissured or jointed, especially when well-graded or clay matrix dominated. They are usually stiff, dense, relatively incompressible soils. Macro, meso and microfabric patterns show high consistency. Sub-horizontal fissuring is due to incremental lodgement and periodic unloading, while subvertical fissures are evidence of ice over-riding and stress relief. Low angle shear failure planes also occur.

Flow till: Consists essentially of melted out super-glacial comminution debris but also occurs due to flow of subglacial meltout tills in subglacial cavities and at the ice margin. It usually contains a wide range of particle sizes. Flow till is frequently interbedded with fluvioglacial material. Orientation of clasts is broadly parallel to the deposition plane and imbricated up flow. The fines also reflect the flow mechanism in the parallelism of clay-silt fractions which produces a locally dense micro-fabric.

Waterlain till: Accumulates on subaqueous surface under a variety of depositional processes and may thus show a wide variety of characters. They vary from rather soft lodgement tills to subaqueous mudflows and to crudely stratified lacustrine clay silts. Stratification, deformation and very diffuse to random clast fabrics are common.

Note: This distinction between superglacial meltout and ablation tills is based on the degree of disturbance of their ice-inherited fabric, including loss of fines. The term ablation till is thus best avoided.

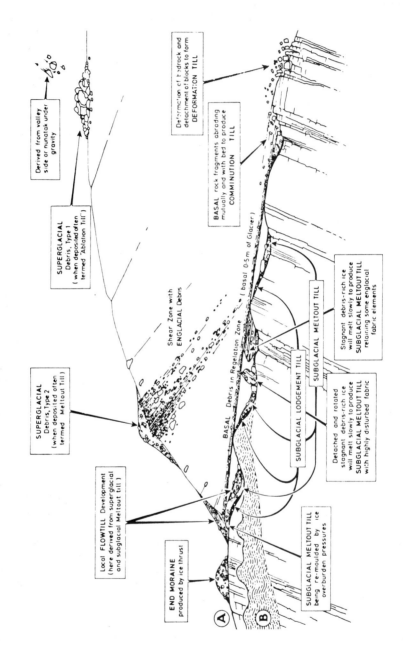

Figure 4.1 Acquisition, transportation and deposition of tills by a glacier. (After McGown and Derbyshire[5]).

a glacial deposit can undergo considerable changes after deposition due to the influence mainly of wind and water.

The particle size distribution and fabric (stone orientation, layering, fissuring and jointing) are among the most significant features as far as the engineering behaviour of a till is concerned. McGown and Derbyshire[5] used the percentage of fines to distinguish granular, well graded and matrix dominated tills, the boundaries being placed at 15 and 45 per cent respectively. The fabric of tills includes features of primary and secondary origins such as folds, thrusts, fissures (macrofabric), disposition of clasts (macro and mesofabric) and the organisation of the matrix. It would seem that distinctive macro and mesofabric patterns characterise flow till, lodgement till and till deformed by thrusting and loading both in terminal moraines and till flutes.

4.2 STRUCTURE OF TILL

Till may possess several structures that affect its strength and permeability. These include microfoliation and preferred orientation of the long axes of the larger rock fragments. The random occurrence in a till of boulders has been referred to as *raisin cake* structure, however, more frequently fragments are crudely aligned in rows rather than in a haphazard manner. Large irregular masses of sand and gravel occur in unpredictable profusion, some of these are interconnected, as a result of meltwater action, but many are isolated. Small distorted pockets of sand and silt have been termed *flame* structures.

Tills are often fissile, tending to split into irregular lenticular flakes, ranging from less than a millimetre thick in clay tills up to several tens of millimetres in sandy tills. Some tills are separated into beds by compact layers of sand a metre or two thick and silt several millimetres thick. Joints commonly extend obliquely through massive, clay tills; some may be caused by shearing and others by desiccation after deposition. Some joints contain thin layers of compact sand.

4.3 GRAIN SIZE DISTRIBUTION

After an investigation of the lithology of tills Krumbein[6] wrote that they possess irregular grain size distributions, being polymodal in aspect. He also concluded that tills were perhaps not so heterogeneous as had previously been supposed and that an individual ice sheet tended to produce till with a fairly well defined frequency distribution. This view was subsequently confirmed by Chryssafopoulos[7].

The grain size distribution has been used to classify tills (Figures 4.2a and b). According to McGown[8] tills are frequently gap graded, the gap generally occurring in the sand fraction. He also noted that large, often very local, variations can occur in the gradings of till which reflect local variations in the formation processes, particularly the comminution processes. The clast size consists principally of rock fragments and composite grains, and presumably was formed by frost action and crushing by ice. Single grains predominate in the matrix. The range in the proportions of coarse and fine fractions in tills dictates the degree to which the properties of the fine fraction influence the properties of the composite soil. The variation in the engineering properties of the fine soil fraction is greater than that of the coarse fraction, and this often tends to dominate the engineering behaviour of the till.

One of the earliest engineering studies of glacial material was made by Legget[9]. The site, near Fergus, Ontario, consisted of fluvio-glacial sands and gravels, and till deposits. Legget found that the specific gravity (now relative

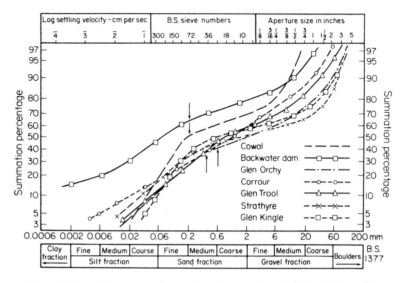

Figure 4.2a Typical gradings of some Scottish morainic soils (After McGown[8])

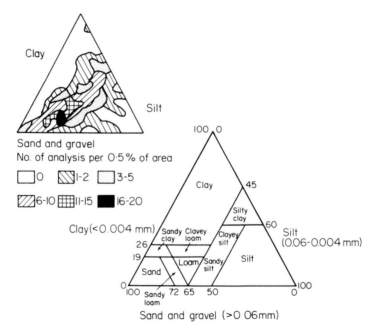

Figure 4.2b Grain size distribution for about 500 tills (After Elson[4])

density of these latter deposits was remarkably uniform, varying from 2.77 to 2.78. These values suggest the presence of fresh minerals in the fine fraction, that is, rock flour rather than clay minerals. Rock flour behaves more like granular material than cohesive and has a low plasticity. Furthermore the investigation revealed that the angle of internal friction of these deposits was always above 30° with a low value of cohesion.

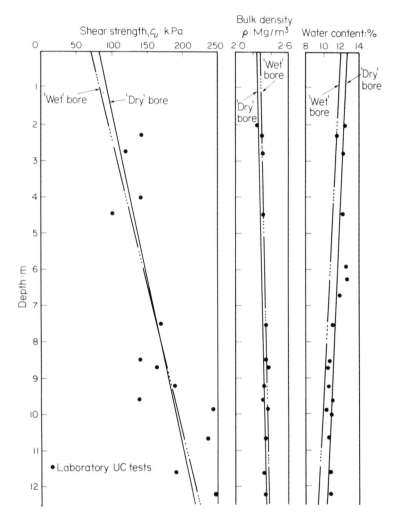

Figure 4.3 'Same day' tests on unconfined compression specimens (samples from 'dry' boreholes) (After McKinley, Tomlinson and Anderson[11])

The consistency limits of tills are dependent upon water content, grain size distribution and the properties of the fine grained fraction. Generally, however, their plasticity index is small and Bernell[10] has shown that the liquid limit of tills decreases with increasing grain size. He added that because of the difficulty of obtaining the plastic limits of Swedish tills by standard methods it had been found more useful to use the liquid and shrinkage limits in classification. The range of water content between these two limits was referred to as the LS difference and Bernell noted that the liquid limit increases linearly with the LS difference. However, McGown[8] dismissed the idea of using the LS difference for classification of tills. McKinley, Tomlinson and Anderson[11] in an investigation of lodgement till showed that plasticity tests on the matrix material placed this till in the CL group, just above the A line on the plasticity chart (e.g. plastic limit = 16%, liquid limit = 28%); the till appeared to be fully saturated. The

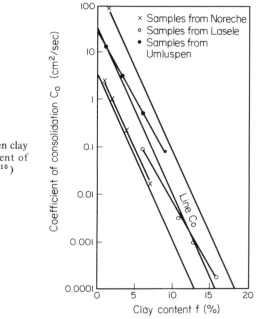

Figure 4.4 Relationship between clay content of till and the coefficient of consolidation (After Bernell[10])

values of moisture content, bulk density and shear strength derived from unconfined compression tests for this till are given in Figure 4.3. This shows that the values of moisture content and bulk density are relatively constant and that the shear strength generally increases with depth.

4.4 COMPRESSIBILITY AND ELASTICITY

The compressibility and consolidation of tills are principally determined by the clay content. For example, the value of compressibility index tends to increase linearly with increasing clay content whilst for moraines of very low clay content, less than 2%, this index remains about constant ($C_c = 0.01$). The relationship between the coefficient of consolidation and clay content is given in Figure 4.4.

Klohn[12] noted that dense, heavily overconsolidated till is relatively incompressible and that when loaded undergoes very little settlement, most of which is elastic. For the average structure such elastic compressions are too small to consider and can therefore be ignored. However, for certain structures they are critical and their magnitude must be estimated prior to construction. Klohn carried out a series of plate loading tests which indicated that the modulus of elasticity of the till deposit concerned was very high, being of the order 1500 MPa. Observations subsequently, taken on all major structures, indicated that settlement occurred almost instantaneously on application of load.

In another survey of dense till, Radhakrishna and Klym[13] found the un-drained shear strength, as obtained by pressuremeter and plate loading tests, to average around 1.6 MPa, while the values from triaxial tests ranged between 0.75 and 1.3 MPa. The average values of the initial modulus of deformation were around 215 MPa which was approximately twice the laboratory value. These differences between field and laboratory results were attributed to stress relief of material on sampling and sampling disturbance. Much lower values of shear strength were found for the Cromer Till by Kazi and Knill[14]; these ranged from 170 to 220 kPa.

4.5 FLUVIO-GLACIAL DEPOSITS

Deposits of stratified drift are often subdivided into two categories, namely, those which develop in contact with the ice – the ice contact deposits; and those which accumulate beyond the limits of the ice, forming in streams, lakes or seas – the proglacial deposits (see Flint[15]).

Outwash fans are deposited by streams which emerge from the snout of a glacier and are composed of sediments ranging in size from coarse sands to boulders. When they are first deposited their porosity may be anything from 25 to 50% and they tend to be very pervious. The finer silt-clay fraction is trans-ported further downstream. Several outwash fans may initially extend from the terminus of a glacier but often they gradually merge to form one deposit. On retreat they may bury the terminal moraine. Kames, kame terraces and eskers are also deposited by melt waters and usually consist of sands and gravels (see Fookes *et al.*[1]).

The most familiar proglacial deposits are varved clays. These deposits accumu-lated in proglacial lakes and are generally characterised by alternating laminae of finer and coarser grain size, each couplet being termed a varve. The thickness of the individual varve is frequently less than 2 mm although much thicker varves have been noted in a few deposits. Generally the coarser layer is of silt size and the finer of clay size.

Taylor *et al*[16] showed that in Devensian clays from Gale Common in Yorkshire, the clay minerals were well orientated around silt grains such that at boundaries between silty partings and matrix the clay minerals tended to show a high degree of orientation parallel to the laminae. Usually very finely comminuted quartz, feldspar and mica form the major part of varved clays rather than clay minerals. For example, the clay mineral content may be as low as 10%, although instances where it has been as high as 70% have been recorded. What is more, Montmorillonite has also been found in varved clays.

Wu[17] noted that the principal constituents of the glacial lake clays found around the shores of the Great Lakes in North America are quartz, kaolinite and illite with small amounts of chlorite and vermiculite. The fabric of these clays varied from well orientated to almost random. He suggested that the latter must involve a flocculent or honeycomb structure. The clays possessed a low strength and high compressibility.

Varved clays tend to be normally consolidated or lightly overconsolidated, although it is usually difficult to make the distinction. In many cases the pre-compression may have been due to ice loading.

The two normally discrete layers formed during the deposition of the varve present an unusual problem in that it may invalidate the normal soil mechanics analyses based on homogeneous soils from being used. As far as the Atterberg limits are concerned, assessment of the liquid and plastic limits of a bulk sample may not yield a representative result. However, Metcalf and Townsend[18] suggested that the maximum possible liquid limit obtained for any particular varved deposit must be that of the clayey portion, whereas the minimum value must be that of the silty portion. Hence they assumed that the maximum and minimum values recorded for any one deposit approximate to the properties of the individual layers. The range of liquid limits for varved clays tends to vary between 30 and 80 whilst that of plastic limit often varies between 15 and 30. These limits, obtained from varved clays in Ontario, allow the material to be classified as inorganic silty clay of medium to high plasticity or compressibility.

In some varved clays in Ontario the natural moisture content would appear to be near the liquid limit. They are consequently soft and have sensitivities generally of the order of 4. Since triaxial and unconfined compression tests tend to give very low strains at failure, around 3%, Metcalf and Townsend presumed that this indicates a structural effect in the varved clays. The average strength reported was about 5.9 MPa, with a range of 3.4 to 7 MPa. The effective stress parameters of apparent cohesion and angle of shearing resistance range from 0.7 to 2.8 MPa, and 22° to 25° respectively.

4.6 QUICK CLAYS

Quick clays are composed of material which is predominantly smaller than 0.002 mm but many deposits seem very poor in actual clay minerals, containing a high proportion of ground down, fine quartz. For example, it has been shown that quick clay from St. Jean Vienney consists of very fine quartz and plagio-clase. Indeed examination of quick clays with the scanning electron microscope has revealed that they do not possess clay based structures, although such work has not lent unequivocal support to the view that non-clay particles govern the physical properties (see Gillott[19]).

Quick clays generally exhibit little plasticity, their plasticity index generally varying between 8 and 12. The most extraordinary property possessed by quick clays is their very high sensitivity. In other words a large proportion of their undisturbed strength is permanently lost following shear. The small fraction of the original strength gained after remoulding may be attributable to the develop-ment of some different form of interparticle bonding. The reason why only a small fraction of the original strength can ever be recovered is because the rate at which it develops is so slow. As an example the Leda Clay is characterised by exceptionally high sensitivity, commonly between 20 and 50, and a high natural moisture content and void ratio, the latter is commonly about 2. It has a low permeability in the range 10^{-10} m/s. The plastic limit is around 25%, with a liquid limit about 60%, and undrained shear strength of 700 kPa. When subjected to sustained load, an undrained triaxial specimen of Leda Clay exhibits a steady time dependent increase in both pore pressure and axial strain. Continu-ing undrained creep may often result in a collapse of the sample after long periods of time have elapsed (Walker[20]).

Quick clays can liquefy on sudden shock. This has been explained by the fact that if quartz particles are small enough, having a very low settling velocity, and if the soil has a high water content, then the solid-liquid transition can be achieved.

The peculiar behaviour and properties of Norwegian quick clays were first explained by Rosenqvist[21], who suggested that they were developed by leaching of clay minerals by fresh water after initial deposition in a marine environment. He proposed that the decrease in salt content of the pore water due to this leaching and diffusion caused the original flocculated structure to become metastable since the change in electrolyte concentration would give rise to particle repulsion rather than attraction. Originally the strength of the soil fabric was dependent on a contribution from the ions in the pore water but when they are leached out by fresh water the strength of the soil depends mainly on the integrity of the fabric. However, Pusch and Arnold[22] disputed the Rosenqvist theory, they having attempted to produce quick behaviour in a specially prepared soil composed largely of illite, following the stages outlined by Rosenqvist, and failed.

The fact that it appears that clay minerals are not quantitatively important in quick clays has led to the development of other theories to explain their peculiar properties. For example, Cabrera and Smalley[23] suggested that these deposits owe their distinctive properties to the predominance of short range interparticle bonding forces which they maintained were characteristic of deposits in which there was an abundance of glacially produced, fine non-clay minerals. In other words they contended that the ice sheets supplied abundant ground quartz in the form of rock flour for the formation of quick clays. Certainly quick clays have a restricted geographical distribution, occurring in certain parts of the northern hemisphere which were subjected to glaciation during Pleistocene times.

4.7 TROPICAL SOILS

Ferruginous and aluminous clay soils are frequent products of weathering in tropical latitudes. They are characterised by the presence of iron and aluminium oxides and hydroxides. The soils may be fine grained or they may contain nodules or concretions. Concretions occur in the matrix where there are higher concentrations of oxides in the soil. More extensive accumulations of oxides give rise to laterite.

Laterite is a residual ferruginous clay-like deposit which generally occurs below a hardened ferruginous crust or hard pan. Ola[24] maintained that in laterites the ratios of silica (SiO_2) to sesquioxides (Fe_2O_3, Al_2O_3) usually are less than 1.33, that those ratios between 1.33 and 2.0 are indicative of lateritic soils, and that those greater than 2.0 are indicative of non-lateritic types.

According to West and Dumbleton[25] lateritic soils form under conditions of strong leaching, that is, high rainfall and temperature, and free drainage. Laterites tend to occur in areas of gentle topography which are not subject to significant erosion.

Laterisation is rapid in tropical regions experiencing periods of heavy rainfall alternating with drier periods. Decomposition of the parent rock involves the removal, by percolating water, of silica (quartz grains resist solution unless they are very fine), lime, magnesia, soda and potash, leading to an enrichment in aluminium and iron oxides. Kaolinite is formed and oxides and hydroxides of iron accumulate, since rapid oxidation does not allow the organic compounds in solution to dissolve iron and thereby for it to be carried away. During the drier periods the water table is lowered. The small amount of iron which has been mobilized in the ferrous state by the ground water is then oxidized, forming hemtatite or, if hydrated, goethite. The movement of the water table leads to the gradual accumulation of iron oxides at a given horizon in the soil profile. A cemented layer of laterite is formed which may be a continuous or honey-combed mass, or nodules may be formed, as in laterite gravel. For example, laterites often have a cellular structure and may consist of ironstone pisolites set in ferruginous cement containing remnant quartz. Concretionary layers are often developed near the surface in lowland areas because of the high water table.

If laterisation proceeds further, as a result of prolonged leaching, then kaolinite is decomposed, the silica being removed in solution and gibbsite remains. Iron compounds may also be removed so that the soil becomes enriched in alumina, a bauxitic soil being developed. Bauxitic clays often have a pisolitic structure.

The colour of lateritic soils varies from very dark red and reddish brown to shades of yellow or pink, depending on the quantity of iron oxide present and its state of hydration.

Gidigasu[26] distinguished three major stages in laterization. The first stage, that of decomposition, is characterised by the physico-chemical breakdown of primary minerals and the release of constituent elements. The second stage involves leaching, under appropriate drainage conditions, of combined silica and bases, and the relative accumulation or enrichment from outside sources of oxides and hydroxides of sesquioxides (mainly Fe_2O_3 and Al_2O_3). The third stage, involves partial or complete dehydration (sometimes involving hardening) of the sesquioxide-rich materials and secondary minerals.

A typical complete laterite profile is:

5. A hard crust very rich in iron (cuirasse or ironstone caprock).
4. A zone rich in free sequioxides, sometimes with kaolinite nodules (laterite proper).
3. A zone of kaolinitic clay material, sometimes with small amounts of mont-morillonite and micas (lithomarge).
2. A decomposed zone of bedrock in which occur relicts of the parent material together with decomposing feldspars.
1. A bedrock usually of igneous material.

The profile indicates how the parent rock is altered, clay minerals developing, with a high proportion of kaolinite. The sesquioxides, especially iron, are con-centrated in the upper part of the profile. Aluminium oxides may replace those of iron in the upper part of the profile and the presence of aluminium may take the form of nodular gibbsite at the surface.

Laterite hardens on exposure to air. Hardening may be due to a change in the hydration of iron and aluminium oxides.

Laterite commonly contains all size fractions from clay to gravel and some-times even larger material (Figure 4.5). Nixon and Skipp[27] provided this range of values for moisture content and consistency limits.

Moisture content   16 to 49%
Liquid limit   33 to 90%
Plasticity index   5 to 59%
Clay fraction   15 to 45%

Usually at or near the surface the liquid limits of laterites do not exceed 60% and the plasticity indices are less than 30%. Consequently laterites are of low to medium plasticity. The activity of laterites may vary between 0.5 and 1.75 (Table 4.2).

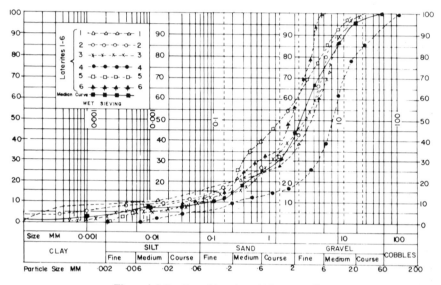

Figure 4.5 Grading of laterites. (After Madu[28])

Table 4.2 SOME PHYSICAL PROPERTIES OF LATERITES AND LATERITIC SOILS

A. From Malaysia (After West and Dumbleton[25]).

	Liquid limit (%)	Plastic limit (%)	Clay fraction (%)	Activity
(1) *Soils overlying basalt*				
Range	46 to 105	31 to 44	9 to 69	0.29 to 0.88
Average	74	39	26	
(2) *Soils overlying granite*				
Range	42 to 107	21 to 34	26 to 54	0.49 to 1.14
Average	71	28	39	
(3) *Soils overlying sedimentary rocks*				
Range	41 to 96	20 to 50	14 to 47	0.46 to 0.76
Average	62	31	39	

B. Other Physical Properties

	Moisture content* (%)	Hygroscopic moisture*	pH value[+]	Void ratio* content (%)
Range	22.8 to 52.5	1.9 to 9.6	1.17 to 1.66	0.8 to 1.9
Average	36.3	4.4	1.34	

	Relative density*	Dry density* (Mg/m³)	Cation exchange capacity[+] (milli-equiv/100 g)	
Range	2.94 to 3.52	4.3 to 5.7	21 to 45	
Average	3.17	5.0	31.6	

*From Tuncer and Lohnes[29]

Lateritic soils, particularly where they are mature, furnish a good bearing stratum. The hardened crust has a low compressibility and therefore settlement is likely to be negligible. In such instances, however, the strength of laterite may decrease with increasing depth. For example, Nixon and Skipp[27] quoted values of shear strength of 90 and 25 kPa, derived from undrained triaxial tests, for laterite samples from a site north of Columbo, Sri Lanka, taken from the surface crust and from a depth of 6 m respectively.

Ola[24] investigated the effects of leaching on lateritic soils. The cementing agents in lateritic soils help to bond the finer particles together to form larger aggregates. However, as a result of leaching these aggregates breakdown, which is shown by the increase in liquid limit after leaching. Moreover removal of cement by leaching gives rise to an increase in compressibility of more than 50%. Again this is due mainly to the destruction of the aggregate structure. Conversely there is a decrease in the coefficient of consolidation by some 20% after leaching (Table 4.3).

Table 4.3 ENGINEERING PROPERTIES OF LATERITIC SOIL BEFORE AND AFTER LEACHING (After Ola[24]).

Property	Before leaching	After leaching
Natural moisture content (%)	14	
Liquid limit (%)	42	53
Plastic limit (%)	25	21
Relative density	2.7	2.5
Angle of shearing resistance, ϕ'	26.5°	18.4°
Cohesion, c' (kPa)	24.1	45.5
Coefficient of compressibility (m²/MN)*	12	15
Coefficient of consolidation (m²/year)*	599	464

*For a pressure of 215 kPa

The change in effective angle of shearing resistance and effective cohesion before and after leaching, is similarly explained. Prior to leaching the larger aggregates in the soil cause it to behave as a coarse-grained, weakly bonded particulate material. The strongly curvilinear form of the Mohr failure envelopes can also be explained as a result of the breakdown of the larger aggregates.

Vertical faces can often be excavated in laterite, up to depths of 6 m, without failing. Deeper cuttings require sloping and drainage at both the top and bottom of the face. Laterite is often a water bearing stratum and heavy pumping may be required for excavations.

Red earths or latosols are residual ferruginous soils in which oxidation readily occurs. Such soils tend to develop in undulating country and most of them appear to have been derived from the first cycle of weathering of the parent material. They differ from laterite in that they behave as a clay and do not possess strong concretions. They do, however, grade into laterite.

The residual red clays of Kenya have been produced by weathering, leaching having removed the more soluble bases and silica, leaving the soil rich in iron oxide (hematite) and hydroxide (goethite) and in alumina. The latter usually occurs in the form of kaolinitic clay minerals or sometimes as gibbsite. Dumbleton[30] found halloysite in these red clays, as did Dixon and Robertson[31].

These soils contain a high percentage of clay size material and have high plastic and liquid limits:

	Clay content (%)	Liquid limit (%)	Plastic limit (%)
Range	63 to 88	76 to 104	34 to 56
Average	80	86	42

Sherwood[32] found that the clay particles in Kenyan red clays were cemented together to form aggregates, the cementing agent being free iron oxide.

Tropical red clays may also be of alluvial origin. Nixon and Skipp[33] found that residual red earths usually plot on or below the 'A' line on the Casagrande plasticity chart whereas alluvial red clays fall above it.

Black clays are typically developed on poorly drained plains in regions with well defined wet and dry seasons, where the annual rainfall is not less than 1250 mm. Generally the clay fraction in these soils exceeds 50%, silty material varying between 20 and 40%, and sand forming the remainder. The organic content is usually less than 2%. The liquid limits of black clays may range between 50 and 100%, with plasticity indices of between 25 and 70%. The shrinkage limit is frequently around 10 to 12% (see Clare[34]) Montmorillonite is commonly present in the clay fraction and is the chief factor determining the behaviour of these clays. For instance, they undergo appreciable volume changes on wetting and drying due to the montmorillonite content. Calcium carbonate occurs sometimes in these black clays, frequently taking the form of concretions (kankar), but it is not an essential characteristic. Usually it constitutes less than one per cent.

The black cotton soil of Nigeria, described by Ola[35], is a highly plastic silty clay which has been formed by the weathering of basaltic rocks, and of shaley and clayey sediments. It may contain up to 70% of montmorillonite, kaolinite and quartz comprising most of the remainder. Shrinkage and swelling of these soils is a problem in many regions of Nigeria which experience alternating wet and dry seasons. The volume changes, however, are confined to an upper critical zone of the soil, which is frequently less than 1.5 m thick. Below this the moisture content remains more or less the same, for instance, around 25%. Ola[36] noted an average linear shrinkage of 8% for some of these soils, with an average swelling pressure of 120 kPa and a maximum of about 240 kPa. He went on to state that in such situations the dead load of a building should be at least 80 kPa to counteract the swelling pressure.

4.8 SOILS IN SEMI-ARID AND ARID REGIONS

Calcareous silty clays are important types of soil in arid and semi-arid regions. They are light to dark brown in colour and are formed by the deposition of clay minerals in saline or lime-rich waters. These soils possess a stiff to hard desiccated clay crust, referred to as duricrust, which may be up to 2 m thick, and which overlies moist soft silty clay. The crust is usually rich in salts precipitated when saline ground water evaporates from the ground surface. Enrichment in lime may lead to the formation of small nodules of limestone in the soil, these are caliche deposits.

Because of the effect of compacting due to rainfall, the upper 25 mm or so of the crust is frequently impervious, beneath which the soil is porous. This impervious surface layer means that run-off is rapid. Uncemented calcareous soils are susceptible to deep erosion, due particularly to sheet flooding. Tomlinson[37] quoted the following range of values for the consistency limits of these soils:

Liquid limit   40 to 59%
Plasticity index   18 to 35%
Shrinkage limit   10 to 15%

Marked swelling and shrinkage is characteristic of calcareous silty clays when alternatively wetted and dried. As a consequence wide, deep cracks may develop in the soil during the dry season.

In arid and semi-arid regions the evaporation of moisture from the surface of the soil may lead to the precipitation of salts in the upper layers. The most commonly precipitated material is calcium carbonate. These caliche deposits are referred to as calcrete. Calcrete occurs where soil drainage is reduced due to long and frequent periods of deficient precipitation and high evapotranspiration. Nevertheless the development of calcrete is inhibited beyond a certain aridity since the low precipitation is unable to dissolve and drain calcium carbonate towards the water table. Consequently in arid climates gypcrete may take the place of calcrete. Climatic fluctuations which, for example, took place in north Africa during Quaternary times, therefore led to alternating calcification and gypsification of soils. Certain calcretes were partially gypsified and elsewhere gypsum formations were covered with calcrete hardpans (see Horta[38]).

The hardened calcrete crust may contain nodules of limestone or be more or less completely cemented (this cement may, of course, have been subjected to differential leaching). In the initial stages of formation calcrete contains less than 40% calcium carbonate and the latter is distributed throughout the soil in a discontinuous manner. At around 40% carbonate content the original colour of the soil is masked by a transition to a whitish colour. As the carbonate content increases it first occurs as scattered concentrations of flakey habit, then as hard conretions. Once it exceeds 60% the concentration becomes continuous (Table 4.2). The calcium carbonate in calcrete profiles decreases from top to base, as generally does the hardness.

Gypcrete is developed in arid zones, that is, where there is less than 100 mm precipitation annually. In the Sahara, aeolian sands and gravels often are encrusted with gypsum deposited from selenitic ground waters. A gypcrete profile may contain three horizons. The upper horizon is rich in gypsified roots and has a banded and/or nodular structure. Beneath this there occurs massive gypcrete-gypsum cemented sands. Massive gypcrete forms above the water table during evporation from the capillary fringe (newly formed gypcrete is hard but it softens with age). At the water table gypsum develops as aggregates of crystals, this is the sand-rose horizon.

Very occasionally in arid areas, enrichment of iron or silica gives rise to ferricrete or silcrete deposits respectively.

Clare[39] described two types of tropical arid soils rich in sodium salts, namely, kabbas and saltmarsh. Both are characterised by their water retentative properties. Kabbas consists of a mixture of partly decomposed coral with sand, clay, organic matter and salt. The salt content in saltmarsh varies up to 40 or 50%, the soil consisting basically of silt with variable amounts of sand and organic material. It occurs in low lying areas that either have a very high water table or are periodically inundated by the sea.

Table 4.4 CLASSIFICATION OF CALCRETE FORMATIONS WITH CONTINUOUS CONCENTRATION OF LIMESTONE (After Horta[38])

Designation	Description	$CaCO_3$ (%)	Thickness	Mutual relationships
Non-foliated calcrete		> 60	Mostly 0.3 to 2–3 m	Under non-foliated calcrete, soft flakes with or without hard nodules are always to be found
Massive	Massive or honeycomb structure			
Nodular	Nodular and honeycomb structure			
Foliated calcrete		> 70	Some centimetres to more than 1 m	Under crusts, there is nearly always non-foliated calcrete. Hardpans exist only at the top of crusts and sometimes totally replace them. Laminated crusts are associated with hardpans and clothe their surface and cracks.
Crusts s.s.	Superimposed and discontinuous sheets of massive or nodular calcrete. The thickness of the sheet is millimetric to centimetric and increases from base to top			
Hardpans	Lithified crust, thickness of the sheets from some centimetres to several decimetres			

For a review of ground conditions in arid regions such as the Middle East, see Fookes[40]

References

1. Fookes, P.G., Gordon, D.L. and Higginbottom, I.E., 'Glacial landforms, their deposits and engineering characteristics.' In *The engineering behaviour of glacial materials, Proc. Symp. Midland Soil Mech. Found. Engng. Soc.*, Birmingham Univ., 18-51 (1975).
2. Boulton, G.S., 'The genesis of glacial tills, a framework for geotechnical interpretation.' In *The engineering behaviour of glacial materials, Proc. Symp. Midland Soil Mech. Found. Engng. Soc.*, Birmingham Univ., 52–59 (1975).
3. Boulton, G.S. and Paul, M.A., 'The influence of genetic processes on some geotechnical properties of glacial tills,' *Q. J. Engng. Geol.*, 9, 159–194 (1976).
4. Elson, J.A., 'Geology of tills,' *Proc. 14th Canadian Conf. Soil Mech.*, Section 3, 5–17 (1961).
5. McGown, A. and Derbyshire, E., 'Genetic influences on the properties of tills,' *Q. J. Engng. Geol.*, 10, 389–410 (1977).
6. Krumbein, W.C., 'Textural and lithologic variations in glacial till,' *J. Geol.*, 41, 382–408 (1933).
7. Chryssafopoulos, H.W.S., 'An example of the use of engineering properties for differentiation of young glacial till sheets,' *Proc. 2nd Pan. Conf. Soil Mech. Found. Engng.*, 2, 35–43 (1963).
8. McGown, A., 'The classification for engineering purposes of tills from moraines and associated landforms,' *Q. J. Engng. Geol.*, 4, 115–130 (1971).

9. Leggett, R., 'An engineering study of glacial drift for an earth dam near Fergus, Ontario,' *Econ. Geol.*, 37, 531–556 (1942).

10. Bernell, L., 'Properties of moraines,' *Proc. 4th Int. Conf. Soil Mech. Found. Engng*, 2, 286–290 (1957).

11. McKinley, D.G., Tomlinson, M.J. and Anderson, W.F., 'Observations on the undrained strength of glacial till,' *Geotechnique*, 24, 503–516 (1974).

12. Klohn, E.J., 'The elastic properties of a dense glacial till deposit', *Can. Geot. J.*, 2, 116–128 (1965).

13. Radhakrishna, H.S. and Klym, T.W., 'Geotechnical properties of a very dense glacial till,' *Can. Geot. J.*, 11, 396–408 (1974).

14. Kazi, A. and Knill, J.L., 'The sedimentation and geotechnical properties of the Cromer till between Happisburgh and Cromer, Norfolk,' *Q. J. Engng. Geol.*, 2, 63–86 (1969).

15. Flint, R.F., *Glacial and Pleistocene geology*, Wiley, New York (1967).

16. Taylor, R.K., Barton, R., Mitchell, J.E. and Cobb, A.E., 'The engineering geology of Devensian deposits underlying PFA lagoons at Gale Common, Yorkshire,' *Q.J. Engng. Geol.*, 9, 195–216 (1976).

17. Wu, T.H., 'Geotechnical properties of glacial lake clays,' *Proc. A.S.C.E., J. Soil Mech. Found. Engng. Div.*, 84, SM3, Paper 1732, 1–36 (1958).

18. Metcalf, J.B. and Townsend, D.L., 'A preliminary study of the geotechnical properties of varied clays as reported in Canadian engineering core records', *Proc. 14th Canadian Conf. Soil Mech.*, Section 13, 203–225 (1961).

19. Gillott, J.E., 'Fabric, composition and properties of sensitive soils from Canada, Alaska and Norway,' *Engng. Geol.*, 14, 149–72 (1979).

20. Walker, L.K., 'Undrained creep in a sensitive clay,' *Geotechnique*, 19, 515–529 (1969).

21. Rosenqvist, I. Th., 'Considerations on the sensitivity of Norwegian quick clays,' *Geotechnique*, 3, 195–200 (1953).

22 Pusch, R. and Arnold, M., 'The sensitivity of artificially sedimented organic free illitic clay,' *Engng. Geol.*, 3, 135–148 (1969).

23. Cabrera, J.G. and Smalley, I.J., 'Quick clays as products of glacial action: a new approach to their nature, geology, distribution and geotechnical properties,' *Engng. Geol.*, 7, 115–133 (1973).

24. Ola, S.A., 'Geotechnical properties and behaviour of stabilized lateritic soils,' *Q. J. Engng. Geol.*, 11, 145–160 (1978).

25. West, G. and Dumbleton, M.J., 'The mineralogy of tropical weathering illustrated by some west Malaysian soils. *Q. J. Engng. Geol.*, 3, 25–40 (1970).

26. Gidigasu, M.D., *Laterite Soil Engineering*, Elsevier, Amsterdam (1976).

27. Nixon, I.K. and Skipp, B.O., 'Airfield construction on overseas soils: Part 5, Laterite. *Proc. Inst. Civ. Engrs.*, 36, Paper no. 6258, 253–275 (1957).

28. Madu, R.M., 'An investigation into the geotechnical and engineering properties of some laterites of eastern Nigeria', *Engng. Geol.*, 11, 101–125 (1977).

29. Tuncer, E.R. and Lohnes, R.A., 'An engineering classification for certain basalt derived lateritic soils,' *Engng. Geol.*, 11, 319–39 (1977).

30. Dumbleton, M.J., 'Origin and mineralogy of African red clays and Keuper Marl,' *Q. J. Engng. Geol.*, 1, 39–46 (1967).

31. Dixon, H.H. and Robertson, R.H.S., 'Some engineering experiences in tropical soils,' *Q. J. Engng. Geol.*, 3, 137–150 (1970).

32. Sherwood, P.T., 'Classification tests on African red clays and Keuper marl. *Q. J. Engng. Geol.*, 1, 47–56 (1967).

33. Nixon, I.K. and Skipp, B.O., Airfield construction on overseas soils: Part 6, Tropical red clays,' *Proc. Inst. Civ. Engrs.*, 36, Paper no. 6258, 275–292 (1957).

34. Clare, K.E., 'Airfield construction on overseas soils' Part 2, Tropical black clays, *Proc. Inst. Civ. Engrs.*, 36, Paper no. 6243, 223–231 (1957).

35. Ola, S.A., 'The geology and engineering properties of the black cotton soils of north eastern Nigeria,' *Engng. Geol.*, 12, 375–391 (1978).

36. Ola, S.A., 'Mineralogical properties of some Nigerian residual soils in relation with building problems,' *Engng. Geol.*, 15, 1–13 (1980).

37. Tomlinson, M.J., 'Airfield construction on overseas soils: Part 3, Saline calcareous soils,' *Proc. Inst. Civ. Engrs.*, 36, paper no. 6239, 232–246 (1957).

38. Horta, J.C. de S.O., 'Calcrete, gypcrete and soil classification,' *Engng. Geol.*, 15, 15–52 (1980).

39. Clare, K.E., 'Airfield construction on overseas soils': Part 1, The formation, classification and characteristics of tropical soils. *Proc. Inst. Civ. Engrs.*, **36**, paper no. 6243, 211–222 (1957).
40. Fookes, P.G., 'Engineering problems associated with ground conditions in the Middle East: inherent ground problems,' *Q. J. Engng. Geol.*, **11**, 33–50 (1978).

Chapter 5

Organic Soils and Fills

5.1 PEAT

Peat is an accumulation of partially decomposed and disintegrated plant remains which have been fossilized under conditions of incomplete aeration and high water content. Physico-chemical and biochemical processes cause this organic material to remain in a state of preservation over a long period of time.

All present day surface deposits of peat have accumulated since the last ice age and therefore have formed during the last 20 000 years. On the other hand, some buried peats may have been developed during inter-glacial periods. Peats have also accumulated in post-glacial lakes and marshes where they are interbedded with silts and muds. Similarly they may be associated with salt marshes. Fen deposits are thought to have developed in relation to the eustatic changes in sea level which occurred after the retreat of the last ice sheets from Britain. The most notable fen deposits in the UK are found south of the Wash. Similar deposits are also found in Suffolk and Somerset. These are areas where expanses of peat interdigitate with wedges of estuarine silt and clay. However, the most familiar type of peat deposit in the UK is probably the blanket bog. These deposits are found on the cool, wet uplands.

Macroscopically peaty material can be divided into three basic groups, namely, amorphous granular, coarse fibrous and fine fibrous peat (see Radforth[1]). The amorphous granular peats have a high colloidal fraction, holding most of their water in an adsorbed rather than a free state, the adsorption occurring around the grain structure. In the other two types the peat is composed of fibres, these usually being woody. In the coarse variety a mesh of second order size exists within the interstices of the first order network whilst in fine fibrous peat the interstices are very small and contain colloidal matter. Most of the water, which these two types of peat contain, is free water. Generally peat deposits are acidic in character, the pH values often varying between 5.5 and 6.5, although some fen peats are neutral or even alkaline.

The ash percentage of peat consists of the mineral residue remaining after its ignition, which is expressed as a fraction of the total dry weight. Ash contents may be as low as 2% in some highly organic peats, or may be as high as 50% as in some peats found on the Yorkshire moors (see Bell[2]). The mineral material is usually quartz sand and silt. In many peat deposits the mineral content increases with depth.

The void ratio of peat ranges between 9, for dense amorphous granular peat, up to 25, for fibrous types with a high content of sphagnum. It usually tends to decrease with depth within a peat deposit. Such high void ratios give rise to phenomenally high water contents. The latter is the most distinctive character- istic of peat. Indeed most of the differences in the physical characteristics of peat are attributable to the amount of moisture present. This varies according to the type of peat, it may be as low as 500% in some amorphous granular varieties whilst, by contrast, percentages exceeding 3000 have been recorded from coarse fibrous varieties. The volumetric shrinkage of peat increases up to a maximum and then remains constant, the volume being reduced almost to the point of complete dehydration. The amount of shrinkage which can occur ranges between 10 and 75% of the original volume of the peat and it can involve reductions in void ratio from over 12 down to about 2.

As would be expected, amorphous granular peat has a higher bulk density than the fibrous types. For instance, in the former it can range up to $1.2 \ Mg/m^3$, whilst in woody fibrous peats it may be half this figure. However, the dry density is a more important engineering property of peat, influencing its be- haviour under load. Hanrahan[3] recorded dry densities of drained peat within the range 65 to $120 \ kg/m^3$. The dry density is influenced by the mineral content and higher values than those quoted can be obtained when peats possess high mineral residues. The relative density of peat has been found to range from as low as 1.1 up to about 1.8, again being influenced by the content of mineral matter.

Because of the variability of peat in the field the value of its permeability as tested in the laboratory can be misleading. Nevertheless Hanrahan[75] showed that the permeability of peat, as determined during consolidation testing, varied according to the loading and length of time involved as follows:

1. Before test: − void ratio = 12; − permeability = 4×10^{-6} m/s.
2. After seven months loading at 55 kPa: − void ratio = 4.5; − permea- bility = 8×10^{-11} m/s.

Thus after seven months of loading, the permeability of the peat was 50 000 times less than it was originally. Miyahawa[5] and Adams[6] have also shown that there is a marked change in the permeability of peat as its volume is reduced under compression. The magnitude of construction pore water pressure is par- ticularly significant in determining the stability of peat. Adams showed that the development of pore pressures in peat beneath embankments was appreciable, in one instance it approached the vertical unit weight of the embankment.

When loaded, peat deposits undergo high deformations but their modulus of deformation tends to increase with increasing load. If peat is very fibrous it appears to suffer indefinite deformation without planes of failure developing. On the other hand failure planes nearly always form in dense amorphous granular peats. Hanrahan and Walsh[7] found that the strain characteristics of peat were independent of the rate of strain and that flow deformation, in their tests, was negligible. Strain often takes place in an erratic fashion in a fibrous peat. This may be due to the different fibres reaching their ultimate strengths at different strain values, the more brittle, woody fibres failing at low strain whilst the non- woody types maintain the overall cohesion of the mass up to much higher strains. The viscous behaviour of peat is generally recognised as being non- Newtonian and the relationship between stress and strain is a function of the void ratio. As the void ratio decreases so the effective viscosity increases and hence a certain value of stress produces a correspondingly smaller value of strain rate.

Apart from its moisture content and dry density the shear strength of a peat deposit appears to be influenced, firstly, by its degree of humification and, secondly, by its mineral content. As both these factors increase so does the shear strength. Conversely the higher the moisture content of peat the lower its shear strength. The dry density is influenced by the effective load to which a deposit of peat has been subjected. As the effective weight of 1 m^3 of drained peat is approximately 45 times that of 1 m^3 of undrained peat the reason for the negligible strength of the latter becomes apparent. Due to its extremely low submerged density, which may be between 15 and 35 kg/m^3, peat is especially prone to rotational failure or failure by spreading, particularly under the action of horizontal seepage forces.

In an undrained bog the unconfined compressive strength is negligible, the peat possessing a consistency approximating to that of a liquid. The strength is increased by drainage to values between 20 and 30 kPa and the modulus of elasticity to between 100 and 140 kPa. According to Hanrahan[8] unconfined compressive strengths of up to 70 kPa are not uncommon in peats consolidated under pavements, a typical modulus of elasticity being 700 kPa (see also Wilson[9]).

5.2 SETTLEMENT AND CONSOLIDATION OF PEAT

Differential and excessive settlement is the principal problem confronting the engineer working on a peaty soil (Figure 5.1). When a load is applied to peat, settlement occurs because of the low lateral resistance offered by the adjacent unloaded peat. Serious shearing stresses are induced even by moderate loads. Worse still, should the loads exceed a given minimum, then settlement may be accompanied by creep, lateral spread, or in extreme cases by rotational slip and upheaval of adjacent ground. At any given time the total settlement in peat due to loading involves settlement with and without volume change. Settlement without volume change is the more serious for it can give rise to the types of failure mentioned. What is more it does not enhance the strength of peat.

Creep does not take place in peat at a constant rate. This is probably due to the increase in density consequent upon consolidation. A good example of the long term behaviour of peat was given by Buisman[10] who cited examples of embankments on peat in the Netherlands in which continuous settlement, linear with the logarithm of time, was recorded for more than 80 years.

When peat is compressed the free pore water is expelled under excess hydrostatic pressure. Since the peat is initially quite pervious and the percentage of pore water is high, the magnitude of settlement is large and this period of initial settlement is short (a matter of days in the field). Adams[6] showed that the magnitude of initial settlement was directly related to peat thickness and applied load. The original void ratio of a peat soil also influences the rate of initial settlement. Excess pore pressure is almost entirely dissipated during this period. Settlement subsequently continues at a much slower rate which is approximately linear with the logarithm of time. This is because the permeability of the peat is significantly reduced due to the large decrease in volume. During this

Figure 5.1 Settlement on peat at Benwick, Hunts. The entrance to the house has settled by about 1 m below pavement level

period the effective consolidating pressure is transferred from the pore water to the solid peat fabric. The latter is compressible and will only sustain a certain proportion of the total effective stress, depending on the thickness of the peat mass.

Adams[11] maintained that the macro- and micro-structure of fibrous peat influences its consolidation. He considered that primary consolidation of such peats took place due to a drainage of water from the macro-structure whilst secondary consolidation was due to the extremely slow drainage of water from the micro-pores into the macro-structure. Because of its higher permeability the rate of primary consolidation of a fine fibrous peat is higher than that of an amorphous granular peat.

Due to the highly viscous water adsorbed around soil particles, amorphous granular peat exhibits a plastic structural resistance to compression and hence has a similar rheological behaviour to that of clay. In this case secondary consolidation is believed to occur as a result of the gradual readjustment of the soil structure to a more stable configuration following the breakdown which occurs during the primary phase due to dissipation of pore pressure. The rate at which this process takes place is controlled by the highly viscous adsorbed water surrounding each soil particle; the colloidal material which the former contains tending to plug the interstices and thereby reduces permeability.

Wilson *et al*[12] suggested that amorphous granular peats exhibit considerable secondary consolidation and therefore settlement. Because of the highly complex structure of such peat they also suggested that it may also exhibit phases of tertiary and quaternary consolidation.

One-dimensional consolidation theories have been developed by Berry and Poskitt[13] for both amorphous granular and fibrous peat. These consider finite strain, decreasing permeability, compressibility and the influence of secondary compression with time. The different mechanisms involved in secondary compression of these two types of peat were found to give similar non-linear rheological models but their effective creep equations were fundamentally different.

That for amorphous granular peat predicts an exponential increase in strain with incremental loading whilst that for fibrous peat predicts a linear increase.

To summarise, the factors which account for the complex manner in which peat consolidates and which therefore mitigate against a precise settlement analysis based on the Terzaghi theory of consolidation, include the abnormally large decrease in permeability which accompanies loading, the decreasing coefficient of compressibility and thixotropy, as well as the surface activity of organic material. What is more it must be remembered that primary and secondary consolidation are empirical divisions of a continuous compression process, both of which occur simultaneously during part of that process.

The quantity of water removed from peat in the later stages of consolidation results in an increase in strength considerably greater than that following the removal of the same quantity during the early stages. What happens to a peat is therefore very largely a function of the structure of the material since this affects the retention and expulsion of water and affords it its strength.

With few exceptions improved drainage has no beneficial effect on the rate of consolidation. This is because efficient drainage only accelerates the completion of primary consolidation which is anyhow completed rapidly.

There is an extremely small increase in the void ratio which follows the reduction of the load on a peat deposit, in other words the voids are not restored to their original value, and the compressibility of preconsolidated peat is greatly reduced. This can be illustrated from the following figures:

Coefficient of volume compressibility of peat for a range of loading from 13.4 to 26.8 kPa

1. Normally loaded $m_v = 12.214 \ m^2/MN$
2. Preconsolidated $m_v = \ \ 0.599 \ m^2/MN$

5.3 FILLS

Because suitable building sites are becoming scarce in urban areas the construction of buildings on fill or made-up ground recently has assumed a greater importance. A wide variety of materials is used for fills including domestic refuse, ashes, slag, clinker, building waste, chemical waste, quarry waste and all types of soils. The extent to which an existing fill will be suitable as a foundation depends largely on its composition and uniformity. In the past the control exercised in placing fill has frequently been insufficient to ensure an adequate and uniform support for structures immediately after placement. Consequently a time interval had to be allowed prior to building so that the material could consolidate under its own weight. Although this may be suitable for small, lightly loaded buildings it is unsatisfactory for more heavily loaded structures which can give rise to substantial settlement.

The stability and potential settlement of foundation structures on fill are largely governed by its density. Therefore random end tipping in thick layers, giving low densities, produces an unsatisfactory condition since excessive and non-uniform settlements are likely to occur under the load of the structure erected. The thickness of fill and site conditions also affect the amount and rate of settlement, the greater the thickness the larger the likely deformation and the increased length of time over which it is likely to take place.

The time taken for a fill to reach a sufficient degree of natural consolidation so that it becomes suitable for a foundation depends on the nature and thickness of the fill, the method of placing and the nature of the underlying ground, especially the ground water conditions. The best materials in this respect are obviously well graded, hard and granular. By contrast, fills containing a large

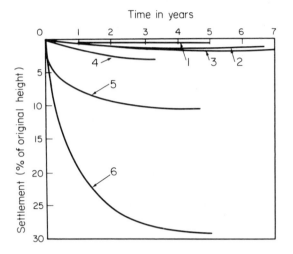

Figure 5.2 Observations of the settlement of various types of fill due to consolidation under its own weight (After Meyerhof[14]). Description of curves: 1. Well-graded sand, well compacted. 2. Rockfill, medium state of compaction. 3. Clay and chalk, lightly compacted. 4. Sand, uncompacted. 5. Clay, uncompacted. 6. Mixed refuse, well compacted

proportion of fine material may take a long while to settle. Generally rock fills will settle 2.5% of their thickness, sandy fills about 5% and cohesive material around 10%. The rate of settlement decreases with time but in some cases it may take 10 to 20 years before movements are reduced within tolerable limits for building foundations. In coarse grained soils the larger part of movement generally occurs within the first two years after the construction of the fill and after five years settlements are usually very small. Indeed Meyerhof[14] showed that most settlement in a fill occurs within the first year of placement and after two years is relatively small (Figure 5.2). Therefore two years generally may be regarded as the lower limit before buildings are constructed on fills consolidating under their own weight.

Fills placed over low lying areas of compressible or weak strata should be considered unsuitable unless tests demonstrate otherwise or the structure can be designed for low bearing capacity and irregular settlement. For example, clay soils beneath fills may undergo a prolonged period of consolidation. Frequently poorly compacted old fills continue to settle for years due to secondary consolidation. Mixed fills which contain materials liable to decay, which may leave voids or involve a risk of spontaneous combustion, afford very variable support and such sites should again, in general, be avoided.

Consequently the material of new fills should be spread uniformly in thin layers and compacted at the optimum moisture content required to produce the maximum density. Properly compacted fills on sound ground can be as good as, or better than, virgin soil (see BRS Digest No. 9[15]). According to Meyerhof[14] the support afforded foundations is generally inadequate if the density of a cohesionless fill is less than 90% of the maximum value given by the standard compaction test. For cohesive material a somewhat higher degree of compaction may be necessary (say around 95%). It has been suggested that if rockfill is compacted to greater than 85% of the solid dry density of the rock then building can commence immediately. For instance, Kilkenny[16] quoted the Expo '67 site in Montreal where a compacted density of 2.3 Mg/m^3 was achieved for the shale fill, as compared with a solid dry density of 2.7 Mg/m^3, and proved satisfactory. Well compacted rock, gravel, sand, shale and clay fills have shown settlements of only 0.5% of their thickness. Quarry wastes frequently have proved satisfactory fills for foundation purposes when properly compacted. Meyerhof indicated that hydraulic sand fills underwent very small settlements if placed above ground water level due to the consolidating influence of the downward percolation of the water. By contrast settlements are large and continue over a long time when clay fill is hydraulically placed.

5.4 BUILDING ON FILL

Settlement distribution over the loaded area of a fill may be very irregular, even under uniform loading conditions. It is necessary therefore to design a structure so that the total, as well as the differential, movements are restricted or can be withstood without damage. Hence, the structure should either be sufficiently rigid to redistribute the loads and thereby reduce relative settlement, or should be relatively flexible to accomodate them without cracks appearing in the structure. Long continuous structures should be avoided; it is best to divide them into sections. Ordinary pad and strip footings are rarely adequate. Wide reinforced strip footings may be adequate where good bearing can be obtained, otherwise a reinforced raft is necessary. If a fill is relatively thin and overlies a firm stratum then a beam and pier or beam and pile foundation may prove an alternative. Settlement observations on structures have revealed that a bearing pressure of 55 kPa is very conservative except on poorly compacted fine grained soils or industrial and domestic wastes consolidating under their own weight.

Waste disposal or sanitary land fills are usually very mixed in composition (Table 5.3) and suffer from continuing organic decomposition and physico-chemical breakdown. Methane and hydrogen sulphide are often produced in the process and accumulations of these gases in pockets in fills have led to explosions. The production of leachate is another problem. Some material such as ashes and industrial wastes may contain sulphates and other products which.are potentially injurious as far as concrete is concerned. The density of waste disposal fills varies from about 120 to 300 kg/m^3 when tipped. After compaction the density may exceed 600 kg/m^3. Moisture contents range from 10 to 50% and the average density index of the solids from 1.7 to 2.5. Settlements are likely to be large and irregular.

According to Sowers[17] the mechanisms responsible for the settlement which occurs in waste disposal fills include: mechanical distortion by the bending, crushing and reorientation of materials which cause a reduction in the void ratio;

Table 5.3 MUNICIPAL WASTE MATERIALS INCORPORATED IN FILLS (AFTER SOWERS, 1973[17])

Material	Characteristics as fill
1. Garbage: Food, waste	Wet. Ferments and decays readily. Compressible, weak
2. Paper, cloth	Dry to damp. Decays and burns. Compressible
3. Graden refuse	Damp. Ferments, decays, burns. Compressible
4. Plastic	Dry. Decay resistant, may burn. Compressible
5. Hollow metal, e.g. drums	Dry. Corrodable and crushable
6. Massive metal	Dry. Slightly corrodable. Rigid
7. Rubber, e.g. tyres	Dry. Resilient, burns, decay resistant. Compressible
8. Glass	Dry. Decay resistant. Crushable and compressible
9. Demolition timber	Dry. Decays and burns. Crushable
10. Building rubble	Damp. Decay resistant. Crushable and erodable
11. Ashes, clinker and chemical wastes	Damp. Compressible, active chemically and partially soluble

ravelling, that is, the transfer of fines into the voids; physico-chemical and bio-chemical changes such as corrosion, combustion and fermentation; and the interactions of these various mechanisms. The initial mechanical settlement of waste disposal fills is rapid and is due to a reduction in the initial void ratio. It takes place with no build up of pore water pressure. Settlement continues due to a combination of secondary compression (material disturbance) and physico-chemical and biochemical action, and Sowers has shown that the settlement-log time relationship is more or less linear. However, the rate of settlement produced by ravelling and combustion is erratic.

Meyerhof[14] referred to a site where industrial and domestic waste had been dumped for some thirty years. When the site was developed only very small movements were recorded where the fill had been in place for many years but on newly filled areas settlements up to 38 mm per month were noted during the early stages of construction. Most of the buildings were of light steel frame construction and sufficiently flexible to accommodate differential movements. Provision was made to jack columns back to their original position if they were subjected to excessive differential settlement. Where large settlements were anticipated pile foundations were used.

Where urban renewal schemes are undertaken it may be necessary to construct buildings on areas covered by rubble fill. In most cases such fills have not been compacted to any appreciable extent and where the rubble has collapsed into old cellars large voids may be present. However, demolition rubble fill is usually comparatively shallow and the most economical method of constructing foundations is either to cut a trench through the fill and backfill it with lean concrete or to clear all the fill beneath the structure and replace it with compacted layers. Deep vibration techniques may prove economical in areas where old cellars make it difficult to operate backacter excavators.

Penman and Godwin[18] noted that maximum rates of settlement occurred immediately after the construction of houses on an old open cast site at Corby which had been backfilled. These settlements decreased to small rates after about four years. The authors suggested that two of the causes of settlement in this fill were creep, which is proportional to log time, and partial inundation. Similar conclusions were reached by Sowers et al[19] for a similar situation. The houses at Corby were constructed twelve years after the fill was placed. The amount of damage which they have suffered is relatively small and is attributable to differential settlement. It is not related to the type of foundation structure used (see Charles *et al*[20]).

Where opencast fills exceed 30 m in depth, because greater settlements may occur, Kilkenny[16] recommended that the minimum time which should elapse before development takes place should be twelve years after restoration is complete. He noted that settlement of opencast backfill appeared to be complete within five to ten years after the operation. For example, comprehensive observations of the opencast restored area at Chibburn, Northumberland, 23 to 38 m in depth, revealed that the ultimate settlement amounted to approximately 1.2% of the fill thickness and that some 50% of the settlement was complete after two years and 75% within five years. In shallow opencast fills, up to 20 m deep, settlements of up to 75 mm have been observed.

5.5 COLLIERY SPOIL

There are two types of colliery discard, namely, coarse and fine. Coarse discard consists of run-of-mine material and reflects the various rock types which are extracted during mining operations. It contains varying amounts of coal which have not been separated by the preparation process. Fine discard consists of either slurry or tailings from the washery, which is pumped into lagoons. Some tips, particularly those with relatively high coal contents, may be partly burnt or burning and this affects their composition and therefore their engineering behaviour.

The moisture content of coarse discard would appear to increase with increasing content of fines, and generally falls within the range 5 to 15%. The range of density index depends on the relative proportions of coal, shale, mudstone and sandstone (Table 2.2). The proportion of coal is of particular importance; the higher the content, the lower the density index. Tip material also shows a wide variation in bulk density and may, in fact, vary within a tip. Low densities are mainly a function of low density index.

The majority of tip material is essentially granular. Often most of it falls within the sand range, but significant proportions of gravel and cobble range may also be present. Due to breakdown, older and surface materials tend to contain a higher proportion of fines than that which occurs within a tip. In coarse discard the liquid and plastic limits are only representative of that fraction passing the 425 μm BS sieve, which frequently is less than 40% of the sample concerned. Nevertheless, the results of these consistency tests suggest a low to medium plasticity whilst in certain instances spoil has proved virtually nonplastic.

As far as effective shear strength of coarse discard is concerned, ϕ' usually varies from 25° to 45°. The angle of shearing resistance, and therefore the strength, increases in spoil which has been burnt. With increasing content of fine coal, on the other hand, the angle of shearing resistance is reduced. The shear strength of colliery spoil, and therefore its stability, is dependent upon the pore pressures developed within it. These are likely to be developed where there is a high proportion of fine material which reduces the permeability below 5×10^{-7} m/s (see Taylor[21]).

Oxidation of pyrite within tip waste is governed by access of air. However, the highly acidic oxidation products which result may be neutralised by alkaline materials in the waste; when this does not happen these chemical changes may give rise to pollution of drainage water. The sulphate content of weathered, unburnt colliery waste is usually high enough to warrant special precautions in the design of concrete structures which may be in contact with the discard or water issuing from it.

Spontaneous combustion of carbonaceous material, frequently aggravated by the oxidation of pyrite, is the most common cause of burning spoil. The problem of combustion has sometimes to be faced when reclaiming old tips (see Bell[22]). The NCB[23] recommends digging out, trenching, blanketing, injection with non-combustible material and water, and water spraying as methods by which spontaneous combustion in spoil can be controlled. Spontaneous combustion may give rise to subsurface cavities in spoil heaps and burnt ashes may also cover zones which are red hot to appreciable depths. When steam comes in contact with red hot carbonaceous material watergas is formed and when the latter is mixed with air it becomes potentially explosive. Explosions may occur when burning spoil heaps are being reworked and a cloud of coal dust is formed near

Table 5.3 EFFECTS OF NOXIOUS GASES. (After the National Coal Board[23])

Gas	Concentration by volume in air p.p.m.	Effect
Carbon monoxide	100	Threshold Limit Value under which it is believed nearly all workers may be repeatedly exposed day after day without adverse effect (T.L.V)
	200	Headache after about 7 hours if resting or after 2 hours if working
	400	Headache and discomfort, with possibility of collapse, after 2 hours at rest or 45 minutes exertion.
	1 200	Palpitation after 30 minutes at rest or 10 minutes exertion
	2 000	Unconsciousness after 30 minutes at rest or 10 minutes exertion
Carbon dioxide	5 000	T.L.V. Lung ventilation slightly increased
	50 000	Breathing is laboured
	90 000	Depression of breathing commences
Hydrogen sulphide	10	T.L.V.
	100	Irritation to eyes and throat: headache
	200	Maximum concentration tolerable for 1 hour
	1 000	Immediate unconsciousness
Sulphur dioxide	1−5	Can be detected by taste at the lower level and by smell at the upper level
	5	T.L.V. Onset or irritation to the nose and throat
	20	Irritation to the eyes
	400	Immediately dangerous to life

Notes. 1. Some gases have a synergic effect, that is, they augment the effects of others and cause a lowering of the concentration at which the symptoms shown in the above table occur. Further, a gas which is not itself toxic may increase the toxicity of one of the toxic gases, for example, by increasing the rate of respiration; strenuous work will have a similar effect.

2. Of the gases listed carbon monoxide is the only one likely to prove a danger to life, as it is the commonest. The others become intolerably unpleasant at concentrations far below the danger level.

the heat surface. If the mixture of coal dust and air is ignited it may explode violently.

Noxious gases are emitted from burning spoil. These include carbon monoxide, carbon dioxide, sulphur dioxide and less frequently hydrogen sulphide. Carbon monoxide is the most dangerous since it cannot be detected by taste, smell or irritation and may be present in potentially lethal concentrations (see Table 5.3). By contrast, sulphur gases are readily detectable in the ways mentioned above and are usually not present in high concentrations.

References

1. Radforth, N.W., 'Suggested classifications of muskeg for the engineer,' *Engineering J. (Canada)*, **35**, 1194–1210 (1952).
2. Bell, F.G., 'Peat: a note and its geotechnical properties,' *Civil Engineering*, 45–49, 49–53 Jan-Feb, 1978.
3. Hanrahan, E.T., 'The mechanical properties of peat with special reference to road construction,' *Trans. Inst. Civ. Engrs.*, Ireland, 78, 179–215 (1952).
4. Hanrahan, E.T., 'An investigation of some physical properties of peat,' *Geotechnique*, **4**, 108–123 (1954).
5. Miyahawa, I., 'Some aspects of road construction over peaty or marshy areas in Hokkaido, with particular reference to filling methods,' *Civ. Engng. Res. Inst.*, Sapporo, Japan (1960).
6. Adams, J.I., 'The engineering behaviour of Canadian muskeg,' *Proc. 6th Int. Conf. Soil Mech. Found. Engng., Montreal*, **1**, 3–7 (1965).
7. Hanrahan, E.T. and Walsh, J.A., 'Investigations of the behaviour of peat under varying conditions of stress and strain,' *Proc. 6th Int. Conf. Soil Mech. Found. Engng., Montreal*, **1**, 226–230 (1965).
8. Hanrahan, E.T., 'A road failure on peat,' *Geotechnique*, **14**, 185–203 (1964).
9. Wilson, N.E., 'The contribution of firbous interlock to the strength of peat,' *Proc. 17th Muskeg Res. Conf.*, ed. G.P. Williams, *Nat. Res. Coun. Canada, Ass. Comm. Geotech. Res., Tech. Mem., No 122*, Ottawa, 5–10 (1978).
10. Buisman, A.S.K., 'Results of long duration settlement tests,' *Proc. 1st Int. Conf. Soil Mech. Found. Engng.*, Cambridge, Mass., **1**, 103–105 (1936).
11. Adams, J.I., 'A comparison of field and laboratory measurement of peat,' *Proc. 9th Muskeg Res. Conf.*, NRC-ACSSM Tech. Memo. 81, 117–135 (1963).
12. Wilson, N.E., Radforth, N.W., Macfarlane, I.C. and Lo, M.B., 'The rates of consolidation for peat,' *Proc. 6th Int. Conf. Soil Mech. Found. Engng.*, Montreal, **1**, 407–412 (1965).
13. Berry, P.L. and Poskitt, T.J., 'The consolidation of peat,' *Geotechnique*, **22**, 27–52 (1972).
14. Meyerhof, G.G., 'Building on fill with special reference to the settlement of a large factory,' *Struct Engng.*, **29**, No. 11, 297–305 (1951).
15. BRS Digest No. 9, *Building on made-up Ground*, Building Research Station, HMSO, Garston, Watford (1967).
16. Kilkenny, W.M., 'A study for the settlement of restored opencast coal sites and their suitability for building development,' *Dept. Civ. Engng, Newcastle Univ., Bull. No. 38* (1968).
17. Sowers, G.E., 'Settlement of waste disposal fills,' *Proc. 8th Int. Conf. Soil Mech. Found. Engng., Moscow*, **2**, 207–212 (1973).
18. Penman, A.D.M. and Godwin, E.W., 'Settlement of experimental houses on land left by opencast mining at Corby.' In *Settlement of Structures*, British Geotechnical Society, Pentech Press, London, 53–61 (1975).
19. Sowers, G.F., Williams, R.C. and Wallace, T.S., 'Compressibility of broken rock and settlement of rockfills,' *Proc. 6th Int. Conf. Soil Mech. Found. Engng., Montreal*, **2**, 561–565 (1965).
20. Charles, J.A., Earle, E.W. and Burford, D., 'Treatment and subsequent performance of

cohesive fill left by opencast mining at Snatchill experimental housing site, Corby, UK. Build. Res. Est., Current Paper No. CP/79/78, Watford (1978).

21. Taylor, R.K., 'Properties of mining wastes with respect to foundations.' In *Foundation Engineering in Difficult Ground* ed. F.G. Bell, Newnes-Butterworths, London, 175– 203 (1978).
22. Bell, F.G., 'Coarse discard from mines', *Civil Engineering,* 37–39 (March 1977).
23. National Coal Board, *Spoil heaps and lagoons,* Technical Handbook, National Coal Board, London (1973).

Chapter 6

Engineering Behaviour of Rock Masses

6.1 THE MECHANICAL BEHAVIOUR OF ROCK

The factors which influence the deformation characteristics and failure of rock masses can be divided into internal and external categories. The internal factors include the inherent properties of the rock itself whilst the external factors are those of its environment at a particular period in time. Regarding internal factors the mineralogical composition and texture are obviously important. However, even more important are the planes of weakness within a rock mass, i.e. the discontinuities and the degree of mineral alteration.

The composition and texture of a rock are governed by its origin. For instance, the olivines, pyroxenes, amphiboles, micas, feldspars and silica minerals are the principal components in igneous rocks. These rocks have solidified from a magma. Solidification involves a varying degree of crystallisation; the greater the length of time involved the greater the development of crystallisation. Hence glassy, fine, medium and coarse grained types of igneous rocks can be distinguished. In metamorphic rocks either partial or complete recrystallisation has been brought about by changing temperature-pressure conditions. Not only are new minerals formed in the solid state but the rocks may develop certain lineation structures, notably cleavage and schistosity. A varying amount of crystallisation is found within the sedimentary rocks. This ranges from almost complete crystallisation as in the case of certain chemical precipitates, to slight such as diagenetic crystallisation in the pores of, for example, certain sandstones.

Few rocks are composed of only one mineral species and even if this happens, the properties of that species vary slightly from mineral to mineral. This in turn is reflected in their physical behaviour. As a consequence few rocks can be regarded as homogeneous, isotropic substances. The size and shape relationships of the component minerals also influence physical behaviour, generally the smaller the grain size the stronger the rock.

One of the most important features of texture as far as physical behaviour, particularly strength, is concerned, is the degree of interlocking of the component grains. Fracture is more likely to take place along grain boundaries than through grains and therefore irregular boundaries make fracture more difficult. The bond between grains in sedimentary rocks is usually provided by the cement and/or matrix, rather than by grains interlocking. The amount and, to a lesser extent, the type of cement/matrix is important, not only influencing strength and elasticity but density, porosity and permeability.

Grain orientation in a particular direction facilitates breakage along that direction. This applies to all fissile rocks whether they are cleaved, schistose, foliated, laminated or thinly bedded. For example, Donath[1] demonstrated that cores cut in Martinsburg Slate at 90° to the cleavage possessed the highest breaking strength whilst those cores cut at 30° exhibited the lowest. Similar experiments were carried out on slate by Hoek[2] who orientated the cleavage planes so as to minimise and maximise their influence. He found that the uni-axial compressive strength in these two directions varied by a factor of approximately four.

Brown *et al*[3] showed that the compressive strength of the Delabole Slate is highly directional, indeed it varies continuously with the angle made by the cleavage planes with the direction of loading (Figure 6.1). They found that even where the cleavage makes high or low angles with the major principal stress direction the mode of failure is mainly influenced by the cleavage. The water

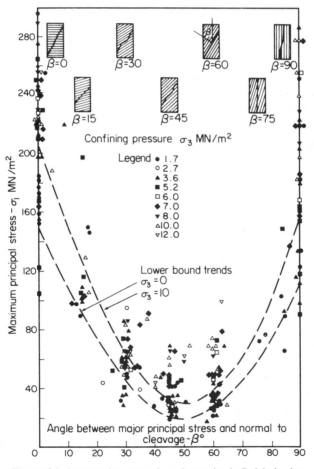

Figure 6.1 Compressive strength amisotrophy in Delabole slate
(After Brown, Richards and Barr[3])

content and surface roughness are the principal factors governing shear strength along the cleavage planes. For example, the average friction angle of smooth wet surfaces was determined as 20.5° which was 9° less than that obtained for the same surfaces when dry. What is more it was found that surface roughness could add up to 40° to the basic friction angles. The degree of surface roughness was shown to vary appreciably according to direction along and character of the cleavage plane concerned, which was reflected in the range of shear strength.

Turning to external factors Griggs[4] noted the changes in rock behaviour with increasing pressure-temperature conditions, simulating increasing depth. He showed that the ultimate strength of the Solenhofen Limestone was increased by 360% under 10 000 atm. High temperatures tend to aid plastic deformation so that it becomes increasingly important with depth. Surface rocks are usually brittle and only some, such as salt, possess plasticity.

Colback and Wiid[5] carried out a number of uniaxial and triaxial compression tests at 8 different moisture contents on quartzitic shale and quartzitic sandstone with porosities of 0.28 and 15% respectively. The tests indicated that

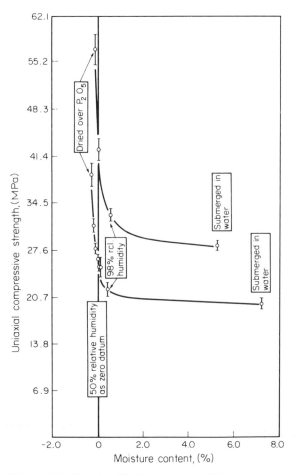

Figure 6.2a Relationship between uniaxial compressive strength and moisture content for quartzitic sandstone

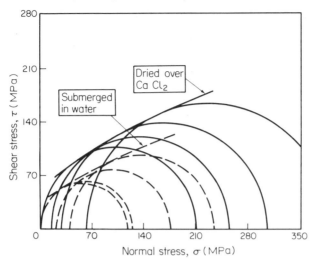

Figure 6.2b Mohr envelopes for quartzitic shale at two moisture
contents (After Colback and Wiid[5])

the compressive strengths of both rocks under saturated conditions were
approximately half what they were under dry conditions (Figure 6.2). From
Figure 6.2 it will be noted that the slopes of the Mohr envelopes are not sensibly
different indicating that the coefficient of internal friction is not significantly
affected by changes in moisture content. Colback and Wiid therefore tentatively
concluded that this reduction in strength was primarily due to a lowering of the
tensile strength which is a function of the molecular cohesive strength of the
material.

Tests on specimens of quartzitic sandstone showed that their uniaxial compressive strength was inversely proportional to the surface tension of the different liquids into which they were placed. As the surface free energy of a solid
submerged in a liquid is a function of the surface tension of the liquid and since
the uniaxial compressive strength is directly related to the uniaxial tensile
strength and this to the molecular cohesive strength, it was postulated that the
influence of the immersion liquid was to reduce the surface free energy of the
rock and hence its strength.

Most strong rocks, such as granite, exhibit little time dependent strain or
creep although creep in evaporitic rocks, notably, salt, may greatly exceed the
instantaneous elastic deformations. Creep in soft chalk should also be given
consideration in civil engineering practice. Although creep deformation usually is
limited at low pressures, it may greatly exceed normal plastic flow when the
pressures approach the limit of rupture. High temperatures also favour an
increase in the rate and extent of creep. From experiments carried out on
Solenhofen Limestone under moderate hydrostatic pressure Robertson[6] concluded that increased hydrostatic pressure caused a very significant decrease in
the primary creep rate per unit stress; and that megascopic fracturing, chiefly
intragranular, decreased rapidly as the confining pressure increased. This conclusion was based upon bulk density measurements and suggested that in limestone under a constant load, fracturing is an important process of primary creep.

It also seems likely that confining pressure might decrease the size, number and propagation of fractures during creep and that it may facilitate their healing when complete unloading has taken place.

6.2 STRENGTH OF ROCK

One of the most popular theories which was proposed to explain shear fractures was advanced by Coulomb[7]. The Coulomb criterion of brittle failure is based upon the idea that shear failure occurs along a surface if the shear stress acting in that plane is high enough to overcome the cohesive strength of the material and the resistance to movement. The latter is equal to the stress normal to the shear surface multiplied by the coefficient of internal friction of the material, whilst the cohesive strength is its inherent shear strength when the stress normal to the shear surface is zero. The Coulomb criterion has been shown to agree with experimental data for rocks in which the relationship between the principal stresses at rupture is, to all intents, linear. If the relationship, however, is non-linear then it is assumed that either the angle of internal shearing resistance is dependent upon pressure or, which is more probable, that the area of the grains in frictional contact increases as the normal pressure increases.

Coulomb's concept was subsequently modified by Mohr[8]. Mohr's hypothesis states that when a rock is subjected to compressive stress shear fracturing occurs parallel to those two equivalent planes for which shearing stress is as large as possible whilst the normal pressure is as small as possible. This assumes that a triaxial state of external stress is applied to a substance and that the maximum external stress is resolved into shear and normal components for any inlined potential shear planes existing in the stressed material. It has been suggested that shear fractures usually enclose an angle of less than $90°$ about the axis of maximum compression because the normal stresses which act across a shear plane are also involved in shear fracturing. The shearing angle is a constant which is governed by the ratio of the ultimate compressional to the ultimate tensile strength. It is therefore characteristic of a particular material at a given temperature and pressure.

In 1920 Griffith[9] claimed that because of the presence of minute cracks or flaws, particularly in surface layers, the measured tensile strengths of most brittle materials are much less than those which would be inferred from the values of the molecular cohesive forces. Although the mean stress throughout a body may be relatively low, local stresses developed in the vicinity of the flaws were assumed to attain values equal to the theoretical strength. Under tensile stress the stress magnification around a flaw is concentrated where the radius of curvature is smallest, in other words at its ends. Hence the tensile stresses which develop around the flaw have the most influence when the tensile stress zone coincides with the zone of minimum radius of curvature. The concentration of stress at the ends of flaws causes them to be enlarged and presumably with time they develop into fractures. Brace[10] showed that the fracture in hard rock was usually initiated in grain boundaries which could be regarded as the inherent flaws required by the Griffith theory. He wrote that as stress was increased prior to fracture grain boundaries at numerous sites in a rock became loosened and that at the instant before fracture it was filled with loosened sections along grain boundaries which had various lengths and orientations. Cracks grew in such sections and ultimately gave rise to fracture.

Griffith's original theory was concerned with brittle fracture produced by applied tensile stress and he based his calculations upon the assumption that the inherent flaw, from which fracture is initiated, could be regarded as an elliptical opening. This simple assumption, however, is not valid in conditions of compressive stress. Accordingly McClintock and Walsh[11] modified the Griffith theory to include the closing of flaws and the development of frictional forces across their surfaces as presumably occurs in compression. Brace[10] noted that in compression tests grain boundaries, although loosened, were in frictional contact at the moment of fracture, and thereby lent support to McClintock and Walsh's modification of the Griffith theory.

Although there is an encouraging agreement between experimental and theoretical results the Griffith theory does not provide a complete description of the mechanism of rock failure for it is only strictly correct when applied to fracture initiation under conditions of static stress. Indeed Hoek[12] contended that it was largely fortuitous that it could be so successfully applied to the prediction of fracture in rock since, once fracture is initiated, its propagation and ultimate failure is a relatively complex process. He further stated that the Griffith theory, when expressed in terms of the stresses at fracture, contributed little to the understanding of rock fracture under dynamic stress conditions, the energy changes associated with fracture or the deformation process of rock.

When the envelope fitted to a set of Mohr circles obtained from low pressure triaxial compression tests on brittle rocks is observed, it is usually represented by a straight line as suggested by the Coulomb equation. As most engineering undertakings involve low confining pressures it is sufficient to assume that the coefficient of friction, which is derived from the slope of the envelope, is a constant. At high confining pressures, however, this assumption is usually erroneous. More importantly it is misleading to assume that the coefficient of friction of 'soft rocks', like shales and mudstones, even at low confining pressures, is a constant. In fact their envelopes are curved. Accordingly it has been concluded that the coefficient of internal friction varies with normal compressive stress. The reason for this has been suggested by Hoek[13] who supposed that it was connected with the interlocking nature of the asperities along the shear plane. He contended that the interlocking depended upon the intimacy of their contact which, in turn, depended upon the amount of normal stress.

One of the most recent empirical laws advanced to describe the shear strength (τ) of intact rock has been given by Barton[14] and is:

$$\tau/\sigma_n = \tan\left[50 \log_{10}\left(\frac{\sigma_1 - \sigma_3}{\sigma_n}\right) + \phi_c\right]$$

where

σ_n is the normal stress;

σ_1 is the axial stress at failure;

σ_3 is the effective confining pressure; and

ϕ_c is equal to $26.6°$.

Barton argued that the effective normal stress mobilised on conjugate shear surfaces was equal to the differential stress $(\sigma_1 - \sigma_3)$ at a critical state. This happens to be the limiting value of the dimensionless ratio $(\sigma_1 - \sigma_3)/\sigma_n$ used in formulating the empirical laws of friction and fracture strength. The critical state for intact rock was defined as the stress condition under which the Mohr envelope of peak shear strength reaches a point of zero gradient. This represents the maximum possible shear strength of the rock. There is a critical effective confining pressure for each rock above which strength cannot increase. The one-dimensional dilation associated with shearing is completely suppressed if the applied stress reaches the level of critical effective confining pressure.

6.3   THE INFLUENCE OF DISCONTINUITIES UPON THE ENGINEERING BEHAVIOUR OF ROCK MASSES

A discontinuity is simply a fracture within a rock mass, the most common discontinuities being joints and bedding planes. Other important discontinuities are cleavage planes, planes of schistosity and fissure zones. In this section the major consideration is given to joints, other discontinuities are dealt with in subsequent sections. Joints are fractures along which little or no displacement has occurred and are present within all rock types. If joints are planar and parallel they may be described as systematic, conversely, when they are irregular they are referred to as non-systematic.

A group of joints which run parallel to each other is termed a joint set whilst two or more joint sets which intersect more or less at a constant angle are referred to as a joint system. When one set of joints is dominant they are called primary joints, the other set being termed secondary. Joints may be classified, on a basis of their size, into master joints, which penetrate several rock horizons and persist for hundreds of metres; smaller joints, which are still well defined structures, are termed major joints; whilst minor joints are related to local structures and do not transcend bedding planes. Lastly, minute fractures occasionally occur in finely bedded sediments and such micro-joints are usually only a few millimetres in size.

Joints may be formed in a number of ways. For example, joints develop within igneous rocks when they cool down, and in wet sediments when they dry out. The most familiar of these are the columnar joints in lava flows, sills and some dykes. Cross joints, longitudinal joints, diagonal joints and flat-lying joints are associated with large granitic intrusions. Sheet or mural joints have a similar orientation to flat-lying joints. When they are closely spaced and well developed they impart a pseudo-stratification to the host rock. It has been noticed that the frequency of sheet jointing is related to depth of overburden, in other words the thinner the rock cover the more pronounced the sheeting. This suggests a connection between removal of overburden by denudation and the development of sheeting. Chapman[15] maintained that other sets of joints developed after the formation of sheet joints and that they were contained between the sheeting surfaces. He showed that the orientation and degree of development of these joints was related to topography.

Price[16], however, postulated that most joints formed after major episodes of folding, that is, they were post-compressional structures. Some spatially restricted small joints associated with folds, such as radial tension joints, are however, probably initiated during folding. He suggested that, after folding, rocks retained residual strain energy and that the associated residual stresses were modified during the succeeding stage of orogenic uplift in such a way as to give rise to tension and shear joints. Indeed joints are frequently associated with folds and faults, but they do not appear to form parallel to other planes of shear failure such as normal and thrust faults. The orientation of joint sets in relation to folds depends upon their size, the type and size of the fold and the thickness and competence of the rocks involved.

The shear strength of a rock mass and its deformability are very much influenced by the discontinuity pattern, its geometry and how well it is developed. Observation of discontinuity spacing, whether in a core or a field exposure, aids the appraisal of rock mass structure. As joints represent surfaces of weakness, the larger and more closely spaced they are, the more influential they become in reducing the effective strength of the rock mass. The nature of the opposing discontinuity surfaces also influences rock mass behaviour as the smoother they are, the more easily can movement take place along them. However, joint surfaces are usually rough and may be slickensided.

Furthermore joints may be open or closed. The degree of opening is of importance in relation to the overall strength and permeability of the rock mass and this often depends largely on the amount of weathering which the rocks have suffered. Some joints may be partially or completely filled. The type and amount of the filling not only influences the effectiveness with which the opposing joint surfaces are bound together, thereby affecting the strength of the rock mass, but it also influences permeability. Mineralisation is frequently associated with joints. This may effectively cement a joint, however, in other cases the mineralising agent may have altered and weakened the rocks along the joint conduit.

Discontinuities in a rock mass reduce its effective shear strength at least in a direction parallel with them. Hence the strength of bedded and jointed rocks is highly anisotropic. Obviously where a load is applied in a direction parallel or subparallel to the discontinuity direction the shear strength depends upon the shearing resistance along that surface. At low normal pressures shearing stresses along a surface with relatively smooth asperities produces a tendency for one block to ride up onto the asperities of the other whereas at high normal pressures shearing takes place through the asperities. When a jointed rock mass undergoes shearing this may be accompanied by dilation especially at low pressures and small shear displacements probably occur as shear stress builds up. Joints offer no resistance to tension whereas they offer high resistance to compression. Nevertheless they may deform under compression if there are crushable asperities, compressible filling material or apertures along the joint surfaces or if the wall rock is altered.

Goodman, Taylor and Brekke[17] considered the application of the finite element method to the problem of jointed rocks. They maintained that for a realistic analysis the jointed rock mass should be treated as an aggregate of massive rock blocks separated by joints rather than as a continuum. In their analysis they distinguished three distinct joint parameters, namely, the unit stiffness across the joint (k_n), the unit stiffness along the joint (k_s) and the shear strength along the joint (S). The value of k_n depends upon the contact area ratio between the joint surfaces, the perpendicular aperture distribution and amplitude, and the relevant properties of the filling materials when present. The controlling factors of the k_s value include the roughness of the joint surfaces which is determined by the distribution, size and inclination of the asperities; and the tangential aperture distribution and amplitude. Lastly the factors which govern the value of S are the friction along the joint, the cohesion due to interlocking, and the strength of the filling material when present.

The moisture content influences all three parameters, however, and the authors regarded the joint water pressures as a separate variable whose effect was analogous to that of pore pressures in soil. Joints with a high stiffness have negligible joint displacements compared with the elastic displacements of the rock blocks whilst those which have a low stiffness value undergo displacements of very much greater magnitude than the elastic displacements of the blocks. A moderate joint stiffness corresponds to displacements of joints to the same extent as the elastic displacements in the blocks. The joint strength was considered as high, moderate or low in accordance with whether the joints played a negligible, participating or dominant role in the strength of the rock mass.

Barton[14, 18] proposed the following expression for deriving the shear strength (τ) along joint surfaces:

$$\tau = \sigma_n \tan \left(JRC \log_{10} \left(\frac{JCS}{\sigma_n} \right) + \phi_b \right)$$

where
   σ_n is the effective normal stress;
   JRC is the joint roughness coefficient;
   JCS is the joint wall compressive strength, and
   ϕ_b is the basic friction angle.

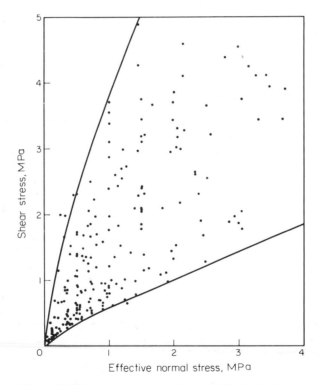

Figure 6.3 The peak shear strength of unfilled joints in rock from the published results of direct shear tests performed both in the laboratory and *in situ* (After Barton[14])

According to Barton the values of the joint roughness coefficient range from 0 to 20, from the smoothest to the roughest surface. The joint wall compressive strength is equal to the unconfined compressive strength of the rock if the joint is unweathered whereas this may be reduced by up to 75% when the walls of the joints are weathered. Both these factors are related since smooth walled joints are less affected by the value of *JCS* since failure of asperities plays a less important role. The smoother the walls of the joints the more significant is the part played by its mineralogy (ϕ_b). The experience gained from rock mechanics indicates that under low effective normal stress levels, such as occur in engineering, the shear strength of joints can vary within relatively wide limits (Figure 6.3) and the maximum effective normal stress acting across joints considered critical for stability lies, according to Barton, in the range 0.1 to 2.0 MPa.

Recent discussions on the influence of joints on the strength of rock masses have been given by Raphael and Goodman[19], Krahn and Morgenstern[20], Bock[21], and Hudson and Priest[22].

6.4 ROCK QUALITY INDICES

Several attempts have been made to relate the numerical intensity of fractures to the quality of unweathered rock masses and to quantify their effect on deformability. For example, the effect of discontinuities in a rock mass can be estimated by comparing the *in situ* compressional wave velocity with the laboratory sonic velocity of an intact core obtained from the same rock mass. The difference in these two velocities is caused by the structural discontinuities which exist in the field. The velocity ratio, V_f/V_1, where V_f and V_1 are the compressional wave velocities of the rock mass *in situ* and of the intact specimen respectively, was first proposed as a quality index by Onodera[23]. For a high quality massive rock with only a few tight joints, the velocity ratio approaches unity. As the degree of jointing and fracturing becomes more severe, the velocity ratio is reduced. The sonic velocity is determined for the core in the laboratory under an axial stress equal to the computed overburden stress at the depth from which the sample was taken, and at a moisture content equivalent to that assumed for the *in situ* rock. The field seismic velocity is preferably determined by uphole or crosshole seismic measurements in drillholes or test adits, since by these measurements it is possible to explore individual homogeneous zones more precisely than by surface refraction surveys.

The concept of rock quality designation (RQD) was introduced by Deere *et al*[24]. Assuming that a consistent drilling standard can be maintained, the percentage of solid core obtained depends on the strength and number of discontinuities in the rock mass. The RQD was defined as the collective length of those core sticks (of NX, i.e. 57.2 mm or larger diameter) in excess of 100 mm, expressed as a percentage of total core length drilled. However, the RQD does not take account of the joint opening and condition, a further drawback being that with fracture spacings greater than 100 mm the quality is excellent irrespective of the actual spacing. This difficulty can be overcome by using the fracture index as suggested by Franklin, Broch and Walton[25], which refers to the frequency with which fractures occur within a rock unit.

The concept of the rock mass factor (*j*) has been recently introduced by Hobbs[26]. He defined the rock mass factor as the ratio of the deformability of a

rock mass within any readily identifiable lithological and structural component
to that of the deformability of the intact rock comprising the component.
Consequently it reflects the effect of discontinuities on the expected performance of intact rock. The value of *j* depends upon the method of assessing the
deformability of the rock mass, and the value beneath an actual foundation will
not necessarily be the same as that determined from even a large scale field test,
particularly on the more massive rocks. Generally the plate loading test, suitably
scaled to the discontinuity spacing, is the best method of determining the field
modulus of deformation. An assessment of rock quality and deformability can
be made by using water injection tests between packers in drillholes. High
hydraulic conductivity in relation to the fracture frequency means correspondingly low *j*-values and vice versa. In simple joint systems, it is theoretically
possible to relate joint opening to hydraulic conductivity. According to Hobbs

Table 6.1 CLASSIFICATION OF ROCK QUALITY IN RELATION TO THE INCIDENCE
OF DISCONTINUITIES

Quality classification	RQD %	Fracture frequency per metre	Velocity ratio V_f/V_1	Mass factor (j)
Very poor	0–25	Over 15	0–0.2	
Poor	25–50	15–8	0.2–0.4	Less than 0.2
Fair	50–75	8–5	0.4–0.6	0.2–0.5
Good	75–90	5–1	0.6–0.8	0.5–0.8
Excellent	90–100	Less than 1	0.8–1.0	0.8–1.0

the greatest difficulties which occur in jointed rock masses in relation to foundation design are experienced in those where the fracture spacing falls within a
range of about 100 to 500 mm, in as much as small variations in fracture spacing
and condition result in exceptionally large changes in the *j*-value. Some of the
classifications of rock quality in relation to discontinuities are given in Table 6.1.

6.5 FAULTS AND FOLDS

The two most important features which are produced when strata are deformed
are faults and folds. Such deformation principally takes place due to movements
along shearing planes. When these are few and large they cause faulting, whilst if
they are small and numerous flexuring and folding results.

Faults are fractures along which adjacent strata have been displaced. The
amount of displacement may vary from only a few millimetres to several
hundred kilometres, for example, the lateral displacement along the San Andreas
fault in California is said to total over 650 km. In some faults the fracture is a
clean break but in others the displacement is not restricted to a simple fracture
but is developed throughout a fault zone, differential movement having occurred
along innumerable closely spaced fractures. If a fault surface is irregular it is
usually grooved in the direction of slip. At the time of their initiation faults are
shear fractures unless they follow pre-existing planes of weakness. A recent
review of the mechanics of fault formation has been given by Murrell[27].

If the movement along a fault has been severe then the rocks involved may
have been crushed, sheared or pulverised (see Sibson[28]). Where shales or clays
have been faulted the fault zone may be occupied by clay gouge (Figure 6.4).

Figure 6.4 Clay gouge occupying the fault zone of the Howick fault in the Limestone Group, Lower Carboniferous, near Howick, Northumberland

Fault breccias, which consist of a jumbled mass of angular fragments containing a high proportion of voids, come into being when more competent rocks are faulted. They are formed by the continued opening of tension gashes where the normal stresses across the fault are low. Accordingly fault breccias are confined to the uppermost region of the crust and are associated with normal faults.

Crush breccias develop when rocks are sheared by a regular pattern of fractures, the individual fragments being bounded by intersecting shear surfaces. If the movements are prolonged then the corners of the fragments may be rounded with the result that crush conglomerates are formed. The material between the lenses forms a fine grained matrix. Crush conglomerates and breccias indicate that strong normal stresses were operational across the fault. These rocks are squeezed as well as stretched and so they are elongated along a plane of flattening which bisects intersecting shear surfaces. Although crush conglomerates may occur in association with any type of fault they most commonly accompany reverse faults.

Movements of even greater intensity are responsible for the occurrence of mylonite along a fault zone, this may be regarded as a micro-breccia and shearing planes may be observed within it. If, because of the severity of the crushing, no individual fragments can be distinguished within the material then it is described as an ultramylonite. Mylonites are principally found in association with low angle thrusts and are thought to be developed at depths where the confining pressures are sufficient to maintain rock coherence.

The ultimate stage in the intensity of movement is reached with the formation of pseudotachylite. This looks like glass and many authorities have suggested that the movements responsible for its formation must have been so severe as to melt the rock material involved and that pseudotachylite was formed upon cooling.

Intraformational shears, that is, zones of shearing parallel to bedding are associated with faulting. Salehy *et al*[29] noted their presence in clays, mudstones and shales of Carboniferous age and that they generally occurred at the contact of such beds with overlying sandstones. They pointed out that these shear zones tended to die out when traced away from the faults concerned and suggested that such zones are probably formed as a result of flexuring of strata adjacent to faults. A shear zone may consist of a single polished or slickensided shear plane, a more complex shear zone may be up to 300 mm in thickness (see Skempton[30]). Intraformational shear zones are not restricted to argillaceous rocks, for instance, they occur in the Chalk. Their presence means that the strength of the rock along the shear zone has been reduced to its residual value.

Shear and tension joints and cleavage are frequently associated with major faults (see Louderback[31]). Shear and tension joints formed along a fault are frequently referred to as feather joints because of their barb-like appearance. Where pinnate shear planes are closely spaced and involve some displacement then fracture cleavage is developed. Slickensides are polished striated surfaces which occur on a fault plane and are caused by the frictional effects generated by its movement.

As a fault is approached the strata involved frequently exhibit flexures which suggest that the beds have been dragged into the fault plane by the frictional resistance generated along it, indeed along some large dip-slip faults the beds may be vertically inclined. A related effect is seen in faulted gneisses and schists where a pre-existing foliation is strongly turned into the fault zone and a secondary foliation results.

In almost all known instances of historic fault breaks the fracturing has occurred along a pre-existing fault. Whilst it seems probable that a given fault would break again at the same location as the last break this cannot be concluded with certainty. However, the likelihood of a new fault interfering with an engineering structure is so remote that it can be reasonably neglected except in unusual situations such as near the tip of the wedge of an active thrust fault.

There is a rough relationship between the length of a fault break and the amount of displacement involved, and both are related to the magnitude of the resultant earthquake. Displacements range from a few millimetres up to 11 m, which was the vertical displacement on the Chedrang fault during the Assam earthquake (1897; $M = 8.7$). The largest horizontal displacement was on the Bogdo fault after the Mongolian earthquake (1957; $M = 8.3$). As expected the longer fault breaks have the greater displacements and generate the larger earthquakes. On the other hand it has been shown by Ambraseys[32] and Bonilla[33] that the smaller the fault displacement the greater the number of observed fault breaks. They also found that for the great majority of fault breaks the maximum displacement was less than 6 m and that the average displacement along the length of the fault was less than 50% of the maximum.

The length of the fault break during a particular earthquake is generally only a fraction of the true length of the fault. Individual fault breaks during simple earthquakes have ranged in length from less than a kilometre to several hundred kilometres. What is more fault breaks do not only occur in association with large and infrequent earthquakes but also occur in association with small shocks and continuous slippage known as fault creep (maximum slip rate on the San Andreas fault is 20 mm per annum).

There is little information on the frequency of breaking along active faults, all that can be said is that some master faults have suffered repeated movements, in some cases recurring in less than 100 years. By contrast much longer intervals totalling many thousands of years have occurred between successive breaks. Therefore because movement has not been recorded in association with a particular fault in an active area it cannot be concluded that the fault is inactive. Earthquakes resulting from displacement and energy release on one fault can

Figure 6.5 Fracture cleavage developed in highly folded Horton Flags, Silurian, near Stainforth, Yorkshire. The inclination of the fracture cleavage is indicated by the near vertical hammer. The other hammer represents the dip of the bedding

sometimes trigger small displacements on other unrelated faults many kilometres distant. Breaks on subsidiary faults have occurred at distances as great as 25 km from the main fault, but with increasing distances from the main fault the amount of displacement decreases. For example, displacements on branch and subsidiary faults located more than 3 km from the main fault break are generally less than 20% of the main fault displacement.

Compressive forces acting in opposing directions cause strata to fold by minute adjustments taking place within and between constituent grains so giving rise to a permanent change in form. The positions of the folds occur where irregularities cause the resolutes of stress to act normal to them. The shape of a fold is influenced by the relative strength of opposing compressive forces and the competency or incompetency of the strata involved. Competent beds transmit stresses whereas incompetent beds behave passively. In fact it is the beds which do not yield to mass flowage which largely determine the shape of a fold. In folding a certain amount of bedding slip takes place. Where competent and incompetent strata are interbedded the movements of the former relative to each other subject the incompetent strata to shearing stress, which may cause small folds, termed drag folds, to develop. Fracture cleavage may be associated with folded competent strata (Figure 6.5).

Shear folding results when minute differential displacements occur along closely spaced shear planes. In shear folding the width of the individual beds along the direction of shear is constant although the beds themselves have been thinned. The apices of the fold are thicker than the flanks if measured normal to the bedding planes. Shear folding may affect folds which have been formed previously.

References

1. Donath, E.A., 'Experimental study of shear failure in anisotropic rocks,' *Bull. Geol. Soc. Am.*, 72, 985–991 (1961).
2. Hoek, E., 'Fracture of anisotropic rock.' *J. S. Af. Inst. Min. Met.*, 64, 510–518 (1964).
3. Brown, E.T., Richards, L.R. and Barr, M.V., 'Shear strength characteristics of the Delabole States,' *Proc. Conf. Rock Engng.*, Newcastle Univ., 1, 33–51 (1977).
4. Griggs, D.T., 'Deformation of rocks under high confining pressures,' *J. Geol.*, 44 (1936).
5. Colback, P.S.B., and Wiid, B.L., 'Influence of moisture content on the compressive strength of rock,' *Symp. Canadian Dept. Min. Tech. Survey.*, Ottawa, 65–83 (1965).
6. Robertson, E.C., 'Creep of Solenhofen Limestone under moderate hydrostatic pressure, in *Rock Deformation,' Geol. Soc. Am.*, Memoir 79, 227–244 (1960).
7. Coulomb, G.A., 'Sur une application des regles de maximus et minimus a quelques problemes relatifs a l'architecture,' *Acad. Roy. des Sci., Mem, de Math, et de Phys. par divers Savans*, 7, 343–382 (1773).
8. Mohr, O., *Abhandlungen aus dem gebiete der technische mechanik*, Ernst und Sohn, Berlin (1882).
9. Griffith, A.A., 'The theory of rupture,' *Proc. 1st Cong. Appl. Mech.*, Delft, 55–70 (1920).
10. Brace, W.F., 'Brittle fracture of rocks. In *Symp. State of Stress in the Earth's Crust*, ed. Judd, W.R, Elsevier, Santa Monica, 111–180 (1964).
11. McClintock, F.A. and Walsh, J.B., 'Friction on Griffith cracks in rocks under pressure,' *Proc. 4th Cong. Appl. Mech.*, 1015–1021 (1962).
12. Hoek, E., 'Rock mechanics – an introduction for the practical engineer', *Min. Mag.*, 114, 236–255 (1966).
13. Hoek, E., 'Brittle fracture of rock' In *Rock Mechanics in Engineering Practice*, ed. Stagg, K.G. and Zienkiewicz, O.C., Wiley, London, 99–124 (1968).
14. Barton, N., 'The shear strength of rock and rock joints,' *Int. J. Rock Mech. Min. Sci.*, 13, 255–279 (1976).
15. Chapman, C.A., 'Control of jointing by topography,' *J. Geol.*, 66, 522–533 (1958).
16. Price, N.J., *Fault and joint development in brittle and semi-brittle rock*, Pergamon, London (1966).
17. Goodman, R.E., Taylor, R.L. and Brekke, T.L., 'A model for the mechanics of jointed rocks,' *Proc. ASCE Soil Mech. Found Engng. Div.*, 94, 637 659 (1968).
18. Barton, N., 'Suggested methods for the quantitative description of discontinuities in rock masses,' *Int. J. Rock Mech. Min. Sci. & Geotech. Abst.*, 15, 319–368 (1978).
19. Raphael, J.M. and Goodman, R.E., 'Strength and deformability of highly fractured rock,' *Proc. ASCE, J. Geotech. Engng. Div.*, 105, GT11, 1285–1300 (1979).
20. Krahn, J. and Morgenstern, N.R., 'The ultimate frictional resistance of rock discontinuities,' *Int. J. Rock Mech. Min. Sci. & Geotech. Abst.*, 16, 127–133 (1979).
21. Bock, H., 'A simple failure criterion for rough joints and compound shear surfaces,' *Engng. Geol.*, 14, 241–54 (1979).
22. Hudson, J.A. and Priest, S.D., 'Discontinuities and rock mass geometry,' *Int. J. Rock Mech. Min. Sci. & Geotech. Abst.*, 16, 339–362 (1979).
23. Onodera, T.F., 'Dynamic investigation of foundation rocks,' *Proc. 5th Symp. Rock Mech. Minnesota*, Pergamon, New York, 517–533 (1963).
24. Deere, D.U., Hendron, A.J., Patton, F.D. and Cording, E.J., 'Design of surface and near-surface construction in rock', *Proc. 8th Symp. Rock Mech., Minnesota*, A.I.M.E., 237–302 (1967).
25. Franklin, J.A., Broch, E. and Walton, G., 'Logging the mechanical character of rock,' *Trans. Inst. Min. Metall.*, 81, Mining Section, A1–9 (1971).
26. Hobbs, N.B., 'Factors affecting the prediction of settlement of structures on rocks with particular reference to the Chalk and Trias.' In *Settlement of Structures*, British Geotechnical Society, Pentech Press, London, 579–610 (1975).
27. Murrell, S.A.F., 'Natural faulting and the mechanics of brittle shear failure,' *J. Geol. Soc.*, 133, 175–190 (1977).
28. Sibson, R.H., 'Fault rocks and fault mechanisms', *J. Geol. Soc.*, 133, 191–214 (1977).
29. Salehy, M.R., Money, M.S. and Dearman, W.R., 'The occurrence and engineering properties of intraformational shears in Carboniferous rocks,' *Proc. Conf. Rock. Engng.*, Newcastle Univ., 1, 311–328 (1977).

30. Skempton, A.W., 'Some observations on tectonic shear zones,' *Proc. 1st Int. Cong. Rock. Mech.,* Lisbon, 1, 329–355 (1966).
31. Louderback, G.D., 'Faults and engineering geology.' In *Application of geology to engineering practice,* Berkey Volume, Am. Geol. Soc., 125–150 (1950).
32. Ambraseys, N.N., 'Maximum intensity of ground movements caused by faulting,' *Proc. 4th World Conf. Earthquake Engng,* Chile, 1, 154–162 (1969).
33. Bonilla, M.G., 'Surface faulting and related effects. In *Earthquake Engineering,* ed. Weigel, R.L., Prentice-Hall, Englewood Cliffs, New Jersey, 47–74 (1970).

Chapter 7

Engineering Classification of Weathered Rocks and Rock Masses

7.1 WEATHERING

Weathering of rocks is brought about by physical disintegration, chemical decomposition and biological activity. The type of weathering which predominates in a region is largely dependent upon climate, which also affects the rate at which weathering proceeds. The latter is also influenced by the stability of the rock mass concerned, which in turn depends upon its mineral composition, texture and porosity, and the incidence of discontinuities within it.

Many rocks were originally formed at high temperatures and pressures and a large part of the weathering process consists of an attempt to reach a new stability under atmospheric conditions. High temperature minerals occur in the ultrabasic and basic igneous rocks. Hence such rocks tend to offer less resistance to weathering than the acid igneous rocks which are largely composed of soda and potash feldspar, quartz and, to a lesser extent, mica. The latter two minerals are particularly stable. Generally coarse grained rocks weather more rapidly than do fine grained types of similar mineral composition.

The degree of interlocking between the constituent minerals is a particularly important textural factor. The closeness of the grains governs the porosity of a rock which in turn determines the amount of water it can hold. Not only are the more porous rocks more susceptible to chemical attack but they are also more prone to frost action. Weathering action is concentrated along discontinuities. For example, the karstic features which occur in thick sequences of limestone are due to solutioning and enlargement of joints and bedding planes; spheroidal weathering, commonly seen in basalts and dolerites, is again caused by weathering along joints (Figures 7.1a and b). The bedding plane spacing in sedimentary rocks affects the rate at which they weather. Intense weathering may be associated with fault zones.

Physical weathering is particularly effective in climatic regions which experience significant changes of diurnal temperature. This does not necessarily imply a wide range of temperature as frost and thaw action can proceed where the range is limited. Alternate freeze-thaw action causes discontinuities to be enlarged and eventually leads to rock shattering. In hot deserts where the daily range of temperature is large the outer layers of rock masses undergo alternate expansion and contraction which eventually leads to them being exfoliated.

Figure 7.1a Gaping Ghyll pothole which descends about 150 m in the Carboniferous Limestone, near Clapham, Yorkshire

Figure 7.1b Spheroidal weathering of basalt lava flows, Tideswell, Derbyshire. In places these lavas are completely decomposed

Chemical weathering leads to mineral alteration and the solution of rocks. Alteration is principally effected by oxidation, hydration, hydrolysis and carbonation whilst solutioning is brought about by acidic or alkaline waters. In dry air, rock decay takes place slowly. However, the presence of moisture hastens the rate tremendously, firstly, because water is itself an effective agent of weathering and, secondly, because it holds in solution substances which react with rock forming minerals. The most important of these substances are free oxygen, carbon dioxide, organic acids and nitrogenous acids. Chemical weathering also aids rock disintegration by weakening the rock fabric and emphasizing any structural weaknesses. When decomposition occurs within a rock the altered material generally occupies a greater volume than the original parental rock material and in the process internal stresses are generated. If swelling occurs in the outer layers of a rock then it causes them to peel off.

Plants and animals play an important role in the breakdown and decay of rocks. Indeed their part in soil formation is of major significance.

The agents of weathering, unlike those of erosion, do not provide transport for the debris from the rock surface. Unless this rock waste is otherwise removed it will eventually act as a protective blanket and so prevent further weathering taking place.

7.2 BREAKDOWN OF SUSCEPTIBLE ROCKS

Certain rocks are, of course, particularly susceptible to breakdown; shales and mudstones fall into this category. The disintegration of shales was studied by Badger, Cummings and Whitmore[1] who concluded that their breakdown was due to the dispersion of colloid material, which appeared to be a general cause of disintegration, and, to a lesser extent, due to air breakage, which only occurs in mechanically weak shales. They found that the variation in disintegration of different shales in water was not usually connected with their total amount of clay colloid or the variation in the types of clay mineral present. It was rather controlled by the type of exchangeable cations associated with clay particles, and by the accessibility of the latter to attack by water which in turn, depended upon the porosity of the shale. Air breakage could assist this process by presenting new surfaces of shale to water. The fracture pattern within mudstones and the lamination of shale aids their disintegration.

Nakano[2] attempted to assess the weatherability of some Japanese mudstones. On being immersed in water these mudstones swelled and underwent a reduction in strength but they did not disintegrate. However, when dried and then wetted again they disintegrated rapidly into small pieces.

Fireclays may disintegrate very rapidly. For instance, Taylor and Spears[3] showed that the Brooch and Park seatearths, of the Sheffield area, after desiccation, broke down in water in less than 30 min, in fact the former was literally explosive. Although the expandable clay content of these two rocks is high and leads to intra-particle swelling, these authors maintained that this alone was not responsible for rapid disintegration. They demonstrated that breakdown could be arrested by removal of air from samples. Thus they concluded that air breakage was a fundamental disintegration mechanism in such rocks. It was suggested that during dry periods evaporation from their surface gives rise to high suction pressures which result in increased shearing resistance. With extreme desiccation most of the voids in a sample of seatearth are filled with air, which, on immersion in water, becomes pressurised by capillary pressures developed in the outer pore spaces. The mineral fabric may then fail along its weakest plane exposing a new surface to the same process.

Swelling of rocks is associated with weathering, clays, shales, mudstones and marls being especially prone. The clay mineral content plays a significant role in these rocks, for example, kaolinite is non-expansive whilst montmorillonite is. Small amounts of swelling have been recorded in some sandstones. The swelling is due principally to the ingress of water therefore the rock must be porous or fractured. Rocks with intact unconfined compressive strengths in excess of 40 MPa are not subject to swelling. When rocks swell they may exert considerable pressures against confining structures, drainage is therefore important. However, it would be better to prevent the ingress of water in the first place. On some sites in swelling rocks, because of the drastic changes which may take place in their character upon wetting, resort to covering the offending strata may be necessary.

The effects of weathering on strengths and deformation moduli are dramatic. For example, a small increase in moisture content or porosity due to weathering can cause large reductions in these parameters, however, the modulus ratio may be only marginally affected. An important consequence of the reduction in modulus is that the joints and fractures have a smaller and smaller influence on the mass factor as the modulus of the intact rock declines, that is, the j-value increases with the increase in the degree of weathering.

A number of tests have been designed to assess certain aspects of weatherability, perhaps one of the most familar being the freeze-thaw test. Freeze-thaw action can quickly break down rocks such as shales and soft chalk. More recently the slake-durability test has been introduced. This test estimates the resistance to wetting and drying of mudstones and shales in particular. Franklin and Chandra[4] found a general qualitative correlation between the slake-durability index, the rate of weathering and the stable slope angle. The most recent engineering classification of rock durability has been proposed by Olivier[5]. This is based on two parameters, the uniaxial compressive strength, and the free swelling coefficient.

7.3 ASSESSMENT OF THE DEGREE OF WEATHERING AND ITS CLASSIFICATION

Several attempts have been made to devise an engineering classification of weathered rock. The problem can be tackled in two ways. Firstly, an attempt can be made to assess the grade of weathering by reference to some simple index test. Such methods provide a quantitative, rather than a qualitative, answer. When coupled with a grading system, this means that the disadvantages inherent in these simple index tests are largely overcome. Hamrol[6] devised a quantitative classification of weatherability in which he first of all distinguished two weathering types. Type I weathering excluded cracking of any kind, whilst in Type II, weathering consisted entirely of cracking. In other words this represents a division between chemical and physical weathering respectively, but such distinction can be extremely difficult to make. In Type I the void ratio increases as weathering progresses, which means that the saturation moisture content increases and the dry density decreases. These two parameters therefore were used as the basis of an index test the numerical value (i_I) of which was expressed as a percentage of water absorbed by a rock in a quick absorption test divided by its dry weight. When considering Type II weathering, Hamrol distinguished between unfilled and filled cracks, the index being

$$i_{II} = (x + y + z) \times 100$$

where x, y and z were the dimensions of the crack along three orthogonal axes. Further indices could be obtained by relating the change in the degree of weathering (j_I and j_{II}) to a given time (Δt), hence

$$j_I = \frac{\Delta i_I}{\Delta t} \quad \text{and} \quad j_{II} = \frac{\Delta i_{II}}{\Delta t}$$

Unfortunately Hamrol gave no scale to his indices so that their meaning in terms of engineering performance is lacking (he did mention that with an $i_I = 10$, a weathered granite would crumble in the fingers, but little else).

Onodera et al[7] also used the number and width of microcracks as an index of the physical weathering of granite. They found a linear relationship between effective porosity and density of microcracks defined as

$$\rho_{cr} = 100 \times (\text{total width of cracks/length of measuring line})$$

They also found that the mechanical strength of granite decreased rapidly as the density of microcracks increased from about 1.5 to 4%.

Lumb[8] defined a quantitative index, Xd, related to the weight ratio of quartz and feldspar in decomposed granite from Hong Kong, as

$$Xd = \frac{N_q - N_{qo}}{1 - N_{qo}}$$

where N_q is the weight ratio of quartz and feldspar in the soil sample, and N_{qo} is the weight ratio of quartz and feldspar in the original rock. For fresh rock $Xd = 0$ while for completely decomposed rock $Xd = 1$.

Table 7.1 STAGES OF WEATHERING OF ROCK MATERIAL IN TERMS OF MICRO-SCOPICAL PROPERTIES (after Irfan and Dearman[9])

Stage 1: No penetration of brown iron-staining. Microcracks are very short, fine, intra-granular and structural. The centres of plagioclases are clouded and slightly sericitized. Altered minerals: <6%; microcrack intensity <0.5%; micropetrographic index >12.

Stage 2: Three substages are recognised depending on the amount of discoloration and type and amount of microfracturing.
   2(i)   The rock is iron-stained only along the joint faces. No penetration of iron-staining.
   2(ii) Penetration of iron-staining (brown) inwards from the joint faces along the microcracks. Formation of simple, branched microcracks; tight and partially stained. Slight alteration of the centres of plagioclases. Occasional staining along quartz-quartz and quartz-feldspar grain boundaries. Grain boundaries are sharp.
   2(iii) More inward penetration of brown iron-staining along microcracks and partial staining of plagioclases. Microfracturing of feldspars and quartz by mainly intragranular, but some transgranular microcracks.
   Unstained core: Altered minerals: 9–12%; microcrack intensity: 0.5–1.0%; micropetrographic index: 9–12.
   Stained rims: Altered minerals: 9–12%; microcrack intensity: 1.0–2.0%; micropetrographic index: 6–9.

Stage 3: Complete discoloration of rock by deep brown iron-staining.
   Partial alteration of plagioclases to sericite and gibbsite (?).
   Formation of single pores in plagioclases due to leaching. Potash feldspars are unaltered. Slight loss of pleochroism and bleaching of biotite. Grain boundaries are tight but stained brown by iron-oxide. The rock fabric is highly microfractured by complex branched, transgranular microcracks. Altered minerals: 12–15%; microcrack intensity: 2.0–5.0%; micropetrographic index: 4–6.

Stage 4: Nearly complete alteration of plagioclase to sericite and gibbsite and formation of nearly opaque areas in plagioclases. Very slight alteration of potash feldspar. Interconnected pores are formed in plagioclase feldspars due to removal and leaching of alteration products. Some solution of silica forming diffused quartz grain boundaries. Some grain boundaries are open. Biotite is bleached to different degrees and altered along grain boundaries. Intense microfacturing of the rock fabric by a complex branched and dendritic pattern of microcracks. The whole of the rock is iron-stained. Altered minerals 15–20%; microcrack intensity: 5.0–10.0%; micropetrographic index: 2–4.

Stage 5: Complete alteration of plagioclases. Potash feldspar is partially altered, but highly microfractured. Biotite is partially and muscovite slightly altered; expansion of biotite. Quartz is reduced in grain size and amount by microfracturing and solution. Almost all the grain boundaries are open. The fabric is intensely microfractured by a dendritic pattern of micro- and macrocracks. Parallel sided, partially filled or clean macrocracks are formed. Highly bleached, highly porous. Rock texture is intact. Altered minerals 20%; microcrack intensity: 10%; micropetrographic index: 2.

As mineral composition and texture influence the physical properties of a rock, petrographic techniques can be used to evaluate successive stages in mineralogical and textural changes brought about by weathering. Accordingly Irfan and Dearman[9] developed a quantitative method of assessing the grade of weathering of granite in terms of its megascopic and microscopic petrography. The megascopic factors included an evaluation of the amount of discolouration, decomposition and disintegration shown by the rock. The microscopic analysis involved assessment of mineral composition and degree of alteration by modal analysis; and a microfracture analysis. The latter involved counting the number of clean and stained microcracks and voids under the microscope in a 10 mm traverse across a thin section. The types of microfracture recognised included stained grain boundaries, open grain boundaries, stained microcracks in quartz and feldspar, infilled microcracks in quartz and feldspar, clean transgranular microcracks crossing the grains, filled or partially infilled microcracks, and pores in plagioclase. The data is used to derive the micropetrographic index:

$$I_p = \frac{\%\text{ sound or primary minerals}}{\%\text{ unsound constituents}}$$

The unsound constituents are the secondary minerals together with microcracks and voids. Irfan and Dearman[9] were able to identify five stages and three substages of weathering in granite (Table 7.1).

Secondly, a classification of weathering can be based on a simple description of the geological character of the rock concerned as seen in the field, the description embodying different grades of weathering which are related to engineering performance (see Moye[10]; Kiersch and Treasher[11]; Knill and Jones[12]). In terms of engineering these classifications have proved the more successful and a recent classification is given in Table 7.2.

7.4 ENGINEERING CLASSIFICATION OF ROCK MASSES

An engineering classification attempts to assess the suitability of a rock mass for a given project and consequently the selection of the parameters for such a classification is of special importance. One of the first attempts at classifying rock masses for engineering purposes was made by Terzaghi[17]. However, this classification has been criticised on the grounds that it was too general to permit an objective evaluation of rock quality and that it provided no quantitative information on the properties of rock masses. Subsequently Deere *et al*[18] introduced the concept of rock quality designation (RQD) as a means of classifying rock masses. The RQD is a more general measure than fracture frequency and is based indirectly on both the degree of fracturing and the amount of weathering in the rock mass.

One of the most recent classification has been advanced by Bieniawski[19]. This initially incorporated the RQD; the unconfined compressive strength; the degree of weathering; the spacing, orientation, separation and continuity of the discontinuities; as well as the ground water flow. The grades of rock quality related to RQD are given in Table 6.1. Unfortunately, however, the RQD takes no account of the spacing, orientation, tightness, roughness of the surface or continuity of discontinuities, or the presence and character of infilling material.

Table 7.2 ENGINEERING GRADE CLASSIFICATION OF WEATHERED ROCK (After Dearman, Fookes and Franklin[13])

Grade	Degree of decomposition	Field recognition (after Fookes and Horswill[14])		Engineering properties	
		Soils (i.e. soft rocks)	Rocks (i.e. hard rocks)	After Little[15]	After Hobbs[16]
VI	Soil	The original soil is completely changed to one of new structure and composition in harmony with existing ground surface conditions	The rock is discoloured and is completely changed to a soil in which the original fabric of the rock is completely destroyed. There is a large volume change	Unsuitable for important foundations. Unstable on slopes when vegetation cover is destroyed and may erode easily unless a hard cap is present. Requires selection before use as fill	In completely weathered rock and residual soil it may be possible to obtain fair quality samples depending upon the parent rock type and the consistency of the product
V	Completely weathered	The soil is discoloured and altered with no trace of original structures	The rock is discoloured and is wholly decomposed and friable, but the original fabric is mainly preserved. The properties of the rock mass depend in part on the nature of the parent rock	Can be excavated by hand or ripping without use of explosives. Unsuitable for foundations of concrete dams or large structures. May be suitable for foundations of earth dams and for fill. Unstable in high cuttings at steep angles. New joint patterns may have formed. Requires erosion protection	Generally the samples will tend to be less disturbed than when taken in the same rock in the highly weathered state. The bearing capacity and settlement characteristics of rock in these extreme states can be assessed using the usual methods for testing soils
IV	Highly weathered*	The soil is mainly altered with occasional small lithorelicts of original soil. Little or no trace of original structures	The rock is discoloured; discontinuities may be open and have discoloured surfaces and the original fabric of the rock near the discontinuities is altered; alteration penetrates	Similar to Grade V. Unlikely to be suitable for foundations of concrete dams. Erratic presence of boulders makes it an unreliable foundation for large structures	In highly weathered rock difficulties will generally be encountered in obtaining undisturbed samples for testing. If samples are obtained the strength and modulus will generally be underestimated, frequently by

Table 7.2 (*continued*)

		deeply inwards, but corestones are still present. The rock mass is partially friable	large margins, even with apparently undisturbed samples. In such rocks *in-situ* tests with either the Menard pressuremeter or the plate should be carried out to determine the bearing capacity and settlement characteristics. The greatest difficulties in assessing bearing capacity and settlement are likely to be encountered in highly weathered rocks, in which the rock fabric becomes increasingly disintegrated or increasingly more plastic
III	Moderately weathered*	The soil is composed of large discoloured lithorelicts of original soil separated by altered material. Alteration penetrates inwards from the surfaces of discontinuities	Excavated with difficulty without the use of explosives. Mostly crushes under bulldozer tracks. Suitable for foundations of small concrete structures and rock fill dams. May be suitable for semipervious fill. Stability in cuttings depends on structural features, especially joint attitudes.
		The rock is discoloured; discontinuities may be open and surfaces will have greater discolouration with the alteration penetrating inwards; the intact rock is noticeably weaker, as determined in the field, than the fresh rock. The rock mass is not friable	In moderately weathered rock the intact modulus, and strength, can be very much lower than in the fresh rock and thus the *j*-value will be higher than in the fresh state, unless the joints and fractures have been opened by erosion or softened by the accumulation of weathering products. The intact modulus and strength can be measured in the laboratory, and the bearing capacity assessed, in the same way as for fresh rock. Triaxial tests may be more appropriate than uniaxial tests, and it would be advisable to adopt conservative values for the factor of safety

*The ratio of the original soil or rock to altered material should be estimated where possible.

Table 7.2 (continued)

		Field recognition (after Fookes and Horswill)		Engineering properties	
Grade	Degree of decomposition	Soils (i.e. soft rocks)	Rocks (i.e. hard rocks)	After Little	After Hobbs
II	Slightly weathered	The material is composed of angular blocks of fresh soil, which may or may not be discoloured. Some altered material starting to penetrate inwards from discontinuities separating blocks	The rock may be slightly discoloured; particularly adjacent to discontinuities which may be open and have slightly discoloured surfaces; the intact rock is not noticeably weaker than the fresh rock	Requires explosives for excavation. Suitable for concrete-dam foundations. Highly permeable through open joints. Often more permeable than the zones above or below. Questionable as concrete aggregate	In faintly and slightly weathered rock it is possible that the j-value, owing to the reduction in stiffness of the joints as a result of penetrative weathering alone, will show a fairly sharp decrease compared with that of the same rock in the fresh state. The intact modulus, by definition, is unaffected by penetrative weathering. The safe bearing capacity is not therefore affected by faint weathering, and may be only slightly affected by slight weathering
I	Fresh rock	The parent soil shows no discolouration, loss of strength or other effects due to weathering	The parent rock shows no discolouration, loss of strength or any other effects due to weathering	Staining indicates water percolation along joints; individual pieces may be loosened by blasting or stress relief and support may be required in tunnels and shafts	

Furthermore in certain circumstances, for instance, when discontinuities are widely spaced or the rock mass is weak, the intact strength of the rock mass has an important bearing on its engineering performance.

Because the unconfined compressive strength and the degree of weathering are two interdependent factors Bieniawski[20] subsequently revised his views and suggested that both factors should be regarded as one parameter, namely, the strength of rock material. He chose a somewhat modified version of Deere's[21] classification of intact strength for his classification:

Very low strength	Under 25 MPa
Low strength	25 to 50 MPa
Medium strength	50 to 100 MPa
High strength	100 to 200 MPa
Very high strength	Over 200 MPa

(cf. Strength classification proposed by Geological Society[22] which provides further subdivision of those rocks with strengths less than 25 MPa and tends to be used in Britain. Bieniawski argued that when assessing importance rating (see below) the strength of such rocks does not contribute to the overall mobility of the rock mass.)

Special caution should be exercised in using this classification in the case of shales and other swelling materials. For instance, Bieniawski suggested that their behaviour under conditions of alternate wetting and drying should be assessed by a slake-durability test (see Franklin and Chandra[4]).

The presence of discontinuities reduces the overall strength of a rock mass and their spacing and orientation govern the degree of such reduction. Hence the spacing and orientation of the discontinuities are of paramount importance as far as the stability of structures in jointed rock masses is concerned. Bieniawski accepted the classification of discontinuity spacing proposed by Deere[23], i.e.

Description	*Spacing of discontinuities*	*Rock mass grading*
Very wide	Over 3 m	Solid
Wide	1 to 3 m	Massive
Moderately close	0.3 to 1 m	Blocky/seamy
Close	50 to 300 mm	Fractured
Very close	Under 50 mm	Crushed

Another revision of Bieniawski's ideas included the continuity and separation of discontinuities which he later[20] grouped together under the heading, condition of discontinuities. This parameter also took account of their surface roughness and the quality of the wall rock. The condition of discontinuities is as important as the discontinuity spacing, for example, tight discontinuities with rough surfaces and no infill have a high strength, whilst, by contrast, open continuous discontinuities facilitate unrestricted flow of ground water. Obviously the condition of discontinuities influences the extent to which the rock material affects the behaviour of a rock mass.

Ground water has an important effect on the behaviour of a jointed rock mass. However, as pore water pressures are of greater significance in foundations than ground water inflow Pells[24] suggested that the pore pressure ratio (r_u), where r_u is defined as the ratio of the pore pressure to the major principal stress, should be included within the classification. This was duly done in the 1974 classification. Bieniawski grouped each of the chosen rock mass parameters into five classes (Table 7.3).

Name of Project:

Site of survey:

Conducted by:

Date:

Structural region	Rock type and origin
	....

Drill core quality R.Q.D.	Weathering
Very good quality 90–100%	Unweathered ....
Good quality 75–90%	Slightly weathered ....
Fair quality 50–75%	Moderately weathered ....
Poor quality 25–50%	Highly weathered ....
Very poor quality < 25%	Completely weathered ....
NOTE: RQD Rock Quality Designation in accordance with the method of	Strength of intact rock material

Ground Water	Designation	Uniaxial compressive strength (MPa)*	Point-load strength index (MPa)
Inflow per 10m	Very high:	Over 200	>8
length litres/min..................	High:	100–200	4–8
or	Medium:	50–100	2–4
Water Pressure kPa	Low:	25–50	1–2
or	Very low:	10–25	<1
General conditions (completely dry,		3–10	
moist only, water under pressure,		1–3	
severe problems)			
....	*1MPa = 1MN/m²		

Figure 7.2
Input Data Form: Engineering classification of jointed rock masses (after Bieniawski[20])

(cont)

Spacing of joints				
		Set 1	Set 2	Set 3	Set 4

		Set 1	Set 2	Set 3	Set 4
Very wide	Over 3m	....	....	....	....
Wide	1–3m	....	....	....	....
Moderately close	0.3–1m	....	....	....	....
Very close	50mm	....	....	....	....

NOTE: These values are obtained from a joint survey and not from borehole logs. Provide data for each joint set.

Strike and dip orientations

Set 1 Strike:............................ (from .... to ....) Dip:............................
                    (average)   (Angle) (Direction)

Set 2 Strike:............................ (from .... to ....) Dip:............................

Set 3 Strike:............................ (from .... to ....) Dip:............................

Set 4 Strike:............................ (from .... to ....) Dip:............................

NOTE: Provide data for each joint set. Refer all directions to magnetic north

Condition of joints

Continuity		Set 1	Set 2	Set 3	Set 4
Not continuous,	no gouge	....	....	....	....
	with gouge	....	....	....	....
Continuous,	no gouge	....	....	....	....
	with gouge	....	....	....	....

Separation	Set 1	Set 2	Set 3	Set 4
Very tight joints: Less than 0.1mm	....	....	....	....
Tight joints 0–1.1mm	....	....	....	....
Moderately open joints 1–5mm	....	....	....	....
Open joints more than 5 mm	....	....	....	....

Roughness	Set 1	Set 2	Set 3	Set 4
Very rough surfaces	....	....	....	....
Rough surfaces	....	....	....	....
Slightly rough surfaces	....	....	....	....
Smooth surfaces	....	....	....	....
Slickensided surfaces	....	....	....	....

Joint wall rock	Set 1	Set 2	Set 3	Set 4
Hardrock	....	....	....	....
Medium hard rock	....	....	....	....
Soft rock	....	....	....	....

NOTE. Provide data for each joint set

Figure 7.2 (continued)   *(cont)*

Figure 7.2 (continued)

Major faults or folds
Describe major faults and folds specifying their locality, nature and orientations
General remarks and additional data
If gouge is present specify its type, thickness, continuity and consistency. Describe waviness of joints. Assess regional stresses.
NOTE: the data on this form constitute the minimum required for engineering design. The geologist, should, however, supply any further information which he considers relevant.

Because these parameters vary in their relative importance from rock mass to rock mass and can contribute individually or collectively to its engineering performance Bienawski used a rating system. In other words a weighted numerical value was given to each parameter, the total rock mass rating being the sum of the weighted values of the individual parameters, the higher the total rating, the better the rock mass conditions (Table 7.3). Bieniawski used a simplified version of the rating system advanced by Wickham *et al*[25]. The ratings given for discontinuity spacings apply to rock masses having three sets of discontinuities. Thus when only one or two sets are present a conservative assessment is attained. In describing a rock mass the class rating should be quoted with the class number, for example, Class 3, rating 68.

Bieniawski provided a summary of the steps which were to be taken when using his classification, they are:

1. The rock mass is divided into structural regions.
2. The input data form (Figure 7.2) is completed for each region and an assessment is made of the input data. For a more recent form see 'The decription of rock masses for engineering purposes', *Q.J. Engng. Geol.*, **10**, 355—88 (1977).

3. Importance ratings are allocated to each parameter as per Table 7.3.
4. The ratings are summed to establish rock mass rating and its class (Table 7.3).
5. The meaning of each rock mass is considered in terms of Table 7.3, as applicable to the given projects.

Barton *et al*[26] proposed the concept of rock mass quality (*Q*) which could be used as a means of rock classification particularly for tunnel support. They defined the rock mass quality in terms of six parameters:

1. The RQD or an equivalent estimate of joint density.
2. The number of joint sets (J_n), which is an important indication of the degree of freedom of a rock mass. The RQD and the number of joint sets provide a crude measure of relative block size.
3. The roughness of the most unfavourable joint set (J_r). The joint roughness and the number of joint sets determine the dilatancy of the rock mass.
4. The degree of alteration or filling of the most unfavourable joint set (J_a). The roughness and degree of alteration of the joint walls or filling materials provides an approximation of the shear strength of the rock mass.
5. The degree of water seepage (J_w)
6. The stress reduction factor (SRF) which accounts for the loading on a tunnel caused either by loosening loads in the case of clay bearing rock masses, or unfavourable stress-strength ratios in the case of massive rock. Squeezing and swelling is also taken account of in the SRF.

They provided a rock mass description and ratings for each of the six parameters which enabled the rock mass quality (**Q**) to be derived from:

$$Q = \frac{RQD}{J_n} \times \frac{J_r}{J_a} \times \frac{J_w}{SRF}$$

This is the most sophisticated method of classifying rocks so far devised. The numerical value of Q ranges from 0.001 for exceptionally poor quality squeezing ground, to 1000 for exceptionally good quality rock which is practically unjointed.

References

1. Badger, C.W., Cummings, A.D. and Whitmore, R.L., 'The disintegration of shale,' *J. Inst. Fuel*, **29**, 417–423 (1956).
2. Nakano, R., 'On weathering and change of properties of Tertiary mudstone related to landslide.' *Soil and Found*, **7**, 1–14 (1967).
3. Taylor, R.K. and Spears, D.A., 'The breakdown of British Coal Measures rocks,' *Int. J. Rock Mech. Min. Sci.*, **7**, 481–501 (1970).
4. Franklin, J.A. and Chandra, R., 'The slake durability test,' *Int. J. Rock Mech. Min. Sci.*, **9**, 325–341 (1972).
5. Olivier, H.J., 'A new engineering-geological rock durability classification,' *Engng. Geol.*, **14**, 255–279 (1979).
6. Hamrol, A., 'A quantitative classification of weathering and weatherability of rocks,' *Proc. 5th Int. Conf. Soil Mech. Found. Engng.*, **2**, 771–773 (1961).

Table 7.3 ENGINEERING CLASSIFICATION OF JOINTED ROCK MASSES (After Bieniawski[19,20])

(a) CLASSIFICATION PARAMETERS AND THEIR RATINGS

1	Strength of intact rock material	Point load strength index (MPa)	> 8	4–8	2–4	1–2	Use of uniaxial compressive test preferred
		Uniaxial compressive strength (MPa)	> 200	100–200	50–100	25–50	< 25
		Rating	10	5	2	1	0
2	Drill core quality RQD		90%–100%	75%–90%	50%–75%	25%–50%	< 25%
		Rating	20	17	14	8	3
3	Spacing of discontinuities		> 3 m	1–3 m	0.3–1 m	50–300 mm	< 50 mm
		Rating	30	25	20	10	5
4	Orientations of discontinuities		Very favourable	Favourable	Fair	Unfavourable	Very unfavourable
		Rating	15	13	10	6	3
5	Condition of discontinuities		Extremely tight. Very rough surfaces Not continuous No separation Hard joint wall rock	Very tight Slightly rough surfaces Separation < 0.1 mm Hard joint wall rock Not continuous	Tight Slightly rough surface Separation < 1 mm No gouge Soft joint wall rock	Open slickensided surfaces OR Gouge < 5 mm OR Joints open 1–5 mm Continuous joints	Very open Soft gouge > 5 mm thick OR Joints open > 5 mm Continuous joints
		Rating	20	15	10	5	0

6 Ground Water

	None	< 25 litres/min Slight	25–125 litres/min Moderate	> 125 litres/min Heavy
Inflow per 10 m tunnel length OR joint water pressure	None			
Ratio = (joint water pressure) / (major principal stress)	0	0.0–0.2	0.2–0.5	> 0.5
General conditions	Completely dry	Moist only (interstitial water)	Water under moderate pressure	Severe water problems
Rating	10	8	5	2

(b) ROCK MASS CLASSES AND THEIR RATINGS

Class No.	I	II	III	IV	V
Description	Very good rock	Good rock	Fair rock	Poor rock	Very poor rock
Rating	100 ← 90	90 ← 70	70 ← 50	50 ← 25	< 25

(c) MEANING OF ROCK MASS CLASSES

Class No.	I	II	III	IV	V
Average stand up time	10 years for 5 m span	6 months for 4 m span	1 week for 3 m span	5 hours for 1.5 m span	10 minutes for 0.5 m span
Cohesion of the rock mass	> 300 kPa	200–300 kPa	150–200 kPa	100–150 kPa	< 100 kPa
Friction angle of the rock mass	< 45°	40°–45°	35°–40°	30°–35°	< 30°
Caveability of ore	Very poor	Will not cave readily. Large fragments	Fair	Will cave readily. Good fragmentation	Very good

7. Onodera, T.F., Yoshinaka, R. and Oda, M., 'Weathering and its relation to mechanical properties of granite,' *Proc. 3rd Cong. Int. Soc. Rock Mech., Denver,* **2A**, 71–78 (1974).
8. Lumb, P., 'The properties of decomposed granite,' *Geotechnique,* **12**, 226–243 (1962).
9. Irfan, T.W. and Dearman, W.R., 'Engineering petrography of a weathered granite,' *Q. J. Engng. Geol.,* **11**, 233–244 (1978).
10. Moye, D.G., 'Engineering geology for the Snowy Mountain scheme,' *Jour. Inst. Engrs. Aust.,* **27**, 287–298 (1955).
11. Kiersch, G.A. and Treasher, R.C., 'Investigations, areal and engineering geology – Folsam Dam Project, Central California,' *Econ. Geol.,* **50**, 271–310 (1955).
12. Knill, J.L. and Jones, K.S., 'The recording and interpretation of geological conditions in the foundations of the Rosieres, Kariba and Latiyan Dams, *Geotechnique,* **15**, 94–124 (1965).
13. Dearman, W.R., Fookes, P.G. and Franklin, J.A., 'Some engineering aspects of weathering with field examples from Dartmoor and elsewhere,' *Q. J. Engng. Geol.,* **3**, 1–24 (1972).
14. Fookes, P.G. and Horswill, P., Discussion on 'The load deformation behaviour of the Middle Chalk at Mundford, Norfolk'. In *In situ investigations in soils and rocks,* British Geotechnical Society, London, 53–57 (1970).
15. Little, A.L., 'The engineering classification of residual tropical soils,' *Proc. 7th Int. Conf. Soil Mech. Found. Engng., Mexico,* **1**, 1–10 (1969).
16. Hobbs, N.B., *Foundations on rock,* Soil Mechanics, Bracknell (1975).
17. Terzaghi, K., 'Introduction to tunnel geology'. In *Rock tunnelling with steel supports* by Procter, R. and White, T. Commercial Shearing and Stamping Co., Youngstown, Ohio, 17–99 (1946).
18. Deere, D.U., Hendron, A.J., Patton, F.D. and Cording, E.J., 'Design of surface and near-surface construction in rock,' *Proc. 8th Symp. Rock Mech., Minnosota, A.I.M.E.,* 237–302 (1967).
19. Bieniawski, Z.T., 'Engineering classification of jointed rock masses,' *Trans. S. Af. Inst. Civ. Engrs.,* **15**, 335–343 (1973).
20. Bieniawski, Z.T., 'Geomechanics classification of rock masses and its application in tunnelling,' *Proc. 3rd Int. Cong. Rock Mech., Denver,* **2**, 27–32 (1974).
21. Deere, D.U., 'Technical description of cores for engineering purposes,' *Rock. Mech. Engng. Geol.,* **1**, 17–22 (1964).
22. Geological Society Engineering Group, 'Working party report on the logging of cores for engineering purposes,' *Q. J. Engng. Geol.,* **3**, 1–24 (1970).
23. Deere, D.U., 'Geological considerations'. In *Rock mechanics in engineering practice,* ed. Stagg, M.G. and Zienkiewiez, O.C., Wiley, London, 1–19 (1968).
24. Pells, P.J.H., 'Discussion: engineering classification of rock jointed masses,' *Trans. S. Af. Inst. Civ. Engrs.,* **16**, 242 (1974).
25. Wickham, G.E., Tiedemann, H.R. and Skinner, E.H., 'Support determination based on geological predictions.' *Proc. 1st N. Am. Tunnelling Conf.,* AIME, New York, 43–64 (1972).
26. Barton, N., Lien, R. and Lunde, J., 'Engineering classification of rock masses for the design of tunnel support,' *Norwegian Geotech. Inst.,* Publ. 106 (1975).

Chapter 8

Engineering Properties of Rocks

8.1 IGNEOUS AND METAMORPHIC ROCKS

The plutonic igneous rocks are characterised by granular texture, massive structure and relatively homogeneous composition. In their unaltered state they are essentially sound and durable with adequate strength for any engineering requirement (Table 8.1). In some instances, however, intrusives may be highly altered, by weathering or hydrothermal attack. Furthermore fissure zones are by no means uncommon in granites. The rock mass may be very much fragmented along such zones, indeed it may be reduced to sand size material (see Terzaghi[1]) and it may have undergone varying degrees of kaolinisation.

Table 8.1 SOME PHYSICAL PROPERTIES OF IGNEOUS AND METAMORPHIC ROCKS

	Relative density	Unconfined compressive strength MPa	Point load strength MPa	Shore scleroscope hardness	Schmidt hammer hardness	Youngs modulus ($\times 10^3$ MPa)
Mount Sorrel Granite	2.68	176.4	11.3	77	54	60.6
Eskdale Granite	2.65	198.3	12.0	80	50	56.6
Dalbeattie Granite	2.67	147.8	10.3	74	69	41.1
Markfieldite	2.68	185.2	11.3	78	66	56.2
Granophyre (Cumbria)	2.65	204.7	14.0	85	52	84.3
Andesite (Somerset)	2.79	204.3	14.8	82	67	77.0
Basalt (Derbyshire)	2.91	321.0	16.9	86	61	93.6
Slate* (North Wales)	2.67	96.4	7.9	41	42	31.2
Slate† (North Wales)		72.3	4.2			
Schist* (Aberdeenshire)	2.66	82.7	7.2	47	31	35.5
Schist†		71.9	5.7			
Gneiss	2.66	162.0	12.7	68	49	46.0
Hornfels (Cumbria)	2.68	303.1	20.8	79	61	109.3

*Tested normal to cleavage or schistocity
†Tested parallel to cleavage or schistocity.

In humid regions valleys carved in granite may be covered with residual soils which extend to depths often in excess of 30 m. Fresh rock may only be exposed in valley bottoms which have actively degrading streams. At such sites it is necessary to determine the extent of weathering and the engineering properties of the weathered products. Generally the weathered product of plutonic rocks has a large clay content although that of granitic rocks is sometimes porous with a permeability comparable to that of medium grained sand.

Joints in plutonic rocks are often quite regular, steeply dipping structures in two or more intersecting sets. Sheet joints tend to be approximately parallel to the topographic surface. The sheet joints introduce a dangerous element of weakness into valley slopes. For example, in a consideration of Mammoth Pool Dam foundations on sheeted granite Terzaghi[2] observed that the most objectionable feature was the sheet joints orientated parallel to the rock surface. In the case of dam foundations such joints, if they remain untreated, may allow the escape of large quantities of water from the reservoir, and this may lead to the development of hydrostatic pressures in the rock downstream which are high enough to dislodge sheets of granite.

Generally speaking the older volcanic deposits do not prove a problem in foundation engineering, ancient lavas having strengths frequently in excess of 200 MPa. But volcanic deposits of geologically recent age at times prove treacherous, particularly if they have to carry heavy loads such as concrete dams. This is because they often represent markedly anisotropic sequences in which lavas, pyroclasts and mud flows are interbedded. Hence foundation problems in volcanic sequences arise because weak beds of ash, tuff and mudstone occur within lava piles which give rise to problems of differential settlement and sliding. In addition weathering during periods of volcanic inactivity may have produced fossil soils, these being of much lower strength. The individual lava flows may be thin and transected by a polygonal pattern of cooling joints. They also may be vesicular or contain pipes, cavities or even tunnels.

Pyroclastics usually give rise to extremely variable foundation conditions due to wide variations in strength, durability and permeability. Their behaviour very much depends upon their degree of induration, for example, many agglomerates have a high enough strength to support heavy loads such as concrete dams and also have a low permeability. By contrast ashes are invariably weak and often highly permeable. One particular hazard concerns ashes, not previously wetted, which are metastable and exhibit a significant decrease in their void ratio on saturation. Tuffs and ashes are frequently prone to sliding. Montmorillonite is not an uncommon constituent in the weathered products of basic ashes.

Slates, phyllites and schists are characterised by textures which have a marked preferred orientation. Platey minerals such as chlorite and mica tend to segregate into almost parallel or subparallel bands alternating with granular minerals such as quartz and feldspar. This preferred alignment of platey minerals accounts for the cleavage and schistocity which typify these metamorphic rocks and means that slate, in particular, is notably fissile. Obviously such rocks are appreciably stronger across, than along the lineation (Table 8.1). The orientation of the planes of cleavage or schistocity in relation to the foundation structure can be significant. Not only does cleavage and schistocity adversely affect the strength of metamorphic rocks, it also makes them more susceptible to decay.

Generally speaking, however, slates, phyllites and schists weather slowly but the areas of regional metamorphism in which they occur have suffered extensive folding so that in places rocks may be fractured and deformed. Some schists, slates and phyllites are variable in quality, some being excellent foundations for heavy structures; others, regardless of the degree of their deformation or weathering, are so poor as to be wholly undesirable. For instance, talc, chlorite and sericite schists are weak rocks containing planes of schistocity only a millimetre or so apart. Some schists become slippery upon weathering and therefore fail under a moderately light load.

The engineering performance of gneiss is usually similar to that of granite. However, some gneisses are strongly foliated which means that they possess a texture with a preferred orientation. Generally this will not significantly affect their engineering behaviour. They may, however, be fissured in places and this can mean trouble. For example, it would appear that fissures opened in the gneiss under the heel of the Malpasset Dam, which eventually led to its failure (see Jaeger[3,4]).

Fresh, thermally metamorphosed rocks such as quartzite and hornfels are very strong and afford excellent foundations. Marble has the same advantages and disadvantages as other carbonate rocks.

8.2 SANDSTONE

Simply defined, a sandstone is an indurated sand, the latter being classified in terms of particle size. In other words a sandstone is a clastic sediment in which mineral grains or rock fragments are bound together with cement and/or matrix. However, there are several fundamental types of sandstone depending on their composition, more particularly the proportions of feldspar, quartz and detrital matrix they contain. For example, quartz arenites contain over 95% quartz whereas greywackes contain 15 to 25% detrital matrix with little or no cement (see Pettijohn, Potter and Siever[5]).

Sandstones may vary from thinly laminated micaceous types to very thickly bedded varieties. Moreover they may be cross-bedded and are invariably jointed. With the exception of shaley sandstone, sandstone is not subject to rapid surface deterioration on exposure.

The dry density and especially the porosity of a sandstone are influenced by the amount of cement and/or matrix material occupying the pores. Usually the density of a sandstone tends to increase with increasing depth below the surface (see Bell[6]).

The compressive strength of a sandstone is influenced by its porosity, amount and type of cement and/or matrix material as well as the composition of the individual grains. Price[7,8] showed that the strength of sandstones with a low porosity (less than 3.5%) was controlled by their quartz content and degree of compaction. In those sandstones with a porosity in excess of 6% he found that there was a reasonably linear relationship between dry compressive strength and porosity, for every 1% increase in porosity the strength decreased by approximately 4%. If cement binds the grains together then a stronger rock is produced than one in which a similar amount of detrital matrix performs the same function. However, the amount of cementing material is more important than the type of cement although if two sandstones are equally well cemented, one having a siliceous, the other a calcareous cement, then the former is the stronger.

For example, ancient quartz arenites in which the voids are almost completely occupied with siliceous material are extremely strong with crushing strengths exceeding 240 MPa. By contrast poorly cemented sandstones may possess crushing strengths less than 3.5 MPa.

The pore water plays a very significant role as far as the compressive strength and deformation characteristics of a sandstone are concerned. For example, the Fell Sandstone and Bunter Sandstone may suffer a reduction of dry compressive strength on saturation of nearly 30 and 60% respectively (Table 8.2).

Moore[9] derived a value of Young's modulus of 1100 MPa for the Bunter Sandstone from long term (up to 18 months) plate load testing. The modulus was found to increase with depth. At the highest loading, 5.6 MPa, settlement did not exceed 4 mm. Creep accounted for 20 to 30% of the total settlement at loads varying between 0.3 and 1.5 MPa, but at 3.0 and 5.6 MPa it was lower. Moore and Jones[10] concluded that at fairly low stresses the Bunter Sandstone, even though weathered near the surface provided a sound foundation. Moreover the rapid reduction in settlement with depth presumably means that simple spread foundation structures may be suitable even for sensitive buildings.

Many sandstones in the valleys excavated in the Millstone Grit series (Namurian) have been fractured by valley bulging or cambering. For example, spectacular valley bulges were recorded in the foundations of the Howden, Derwent and

Table 8.2 SOME PHYSICAL PROPERTIES OF ARENACEOUS SEDIMENTARY ROCKS

	Fell Sandstone (Rothbury)	Chatsworth Grit (Stanton in the Peak)	Bunter Sandstone (Edwinstowe)	Keuper Waterstones (Edwinstowe)	Horton Flags (Helwith Bridge)	Bronllwyn Grit (Llanberis)
Relative density	2.69	2.69	2.68	2.73	2.70	2.71
Dry density (Mg/m^3)	2.25	2.11	1.87	2.26	2.62	2.63
Porosity	9.8	14.6	25.7	10.1	2.9	1.8
Dry unconfined compressive strength (MPa)	74.1	39.2	11.6	42.0	194.8	197.5
Saturated unconfined compressive strength (MPa)	52.8	24.3	4.8	28.6	179.6	190.7
Point load strength (MPa)	4.4	2.2	0.7	2.3	10.1	7.4
Scleroscope hardness	42	34	18	28	67	88
Schmidt hardness	37	28	10	21	62	54
Young's modulus $(\times 10^3$ MPa)	32.7	25.8	6.4	21.3	67.4	51.1
Permeability $(\times 10^{-9}$ m/s)	1740	1960	3500	22.4	–	–

Ladybower Dams. In the latter the folding was present to a depth of almost 60 m (see Hill[11]). A further consequence of valley bulges is the opening up of tension fissures in sandstones forming the valley sides. These fissures run parallel to the valley and may be up to 250 mm wide close to the valley side, but they become progressively narrower and finally disappear when followed into the hillside. Valley bottom and valley side disturbance appear to be the result of a number of factors including stress relief, moisture up-take by underlying shales, artesian water pressures, valley notch concentration of stresses or frozen ground conditions.

Frequently thin beds of sandstone and shale are interbedded. Foundations on such sequences may give rise to problems of shear, settlement, and rebound, the magnitude of these factors depending upon the character of the shales. In some cases this even accentuates the undesirable properties of the shale by permitting access of water to the shale-sandstone contacts. Contact seepage may weaken shale surfaces and cause slides in dipping formations.

8.3 CARBONATE ROCKS

Carbonate rocks contain more than 50% of carbonate minerals, amongst which calcite and/or dolomite predominate. Normally the term limestone is used for those rocks in which the carbonate fraction is composed principally of calcite and the term dolostone is reserved for those rocks in which dolomite accounts for more than half the carbonate fraction. Chalk is a rather peculiar type of soft, remarkably pure limestone which is characteristically developed in the Upper Cretaceous.

Fookes and Higginbottom[12] devised a classification of limestones for engineering purposes. They chose mineral composition, origin, grain size and degree of induration as the basis of their classification (Figure 8.1).

Representative values of some physical properties of carbonate rocks are listed in Table 8.3. It can be seen that, generally, the density of these rocks increases with age, whilst the porosity is reduced. Diagenetic processes mainly account for the lower porosities of the Carboniferous and Magnesion Limestones in particular. On the other hand the high porosity values of the Upper Chalk may be due to the presence of hollow tests and the complex shapes of the constituent particles. What is more, the Upper Chalk is very poorly cemented and has not suffered the same degree of pre-consolidation loading as the Middle and Lower Chalk.

Age often has an influence on the strength and deformation characteristics of carbonate and other sedimentary rocks. From Table 8.3 it can be seen that Carboniferous Limestone is generally very strong, conversely the Bath Stone (Great Oolite, Jurassic) is only just moderately strong. Similarly the oldest limestones tend to have the highest values of Young's modulus.

Thick-bedded, horizontally lying limestones relatively free from solution cavities afford excellent foundations. For example, intact samples of Carboniferous Limestone may have unconfined compressive strengths greater than 100 MPa. On the other hand thin bedded, highly folded or cavernous limestones are likely to present serious foundation problems. A possibility of sliding may exist in thinly bedded, folded sequences. Similarly if beds are separated by layers of clay or shale these, especially when inclined, may serve as sliding planes and result in failure.

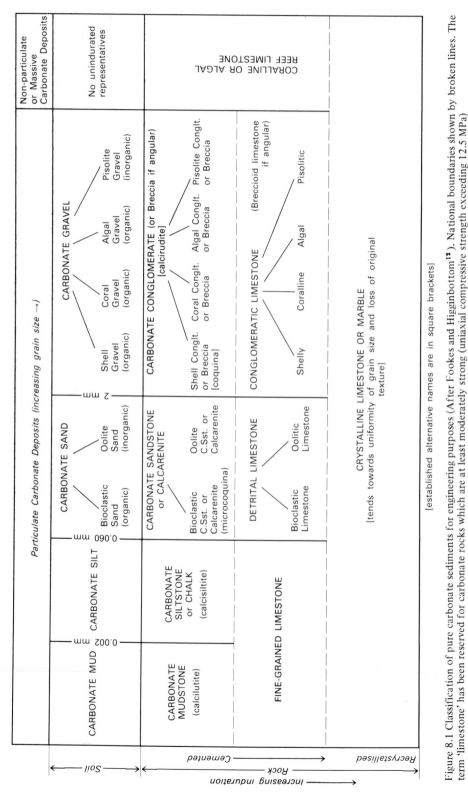

Figure 8.1 Classification of pure carbonate sediments for engineering purposes (After Fookes and Higginbottom[12]). National boundaries shown by broken lines. The term 'limestone' has been reserved for carbonate rocks which are at least moderately strong (uniaxial compressive strength exceeding 12.5 MPa)

Table 8.3 SOME PHYSICAL PROPERTIES OF CARBONATE ROCKS

	Carbon- iferous Limestone (Buxton)	Magnesium Limestone (Anston)	Ancaster Freestone (Ancaster)	Bath Stone (Corsham)	Middle Chalk (Hillington)	Upper Chalk (North- fleet)
Relative density	2.71	2.83	3.70	2.71	2.70	2.69
Dry density (Mg/m³)	2.58	2.51	2.27	2.30	2.16	1.49
Porosity (%)	2.9	10.4	14.1	15.6	19.8	41.7
Dry unconfined compressive strength MPa	106.2	54.6	28.4	15.6	27.2	5.5
Saturated uncon- fined compres- sive strength MPa	83.9	36.6	16.8	9.3	12.3	1.7
Point load strength (MPa)	3.5	2.7	1.9	0.9	0.4	–
Scleroscope hardness	53	43	38	23	17	6
Schmidt hardness	51	35	30	15	20	9
Young's modulus (× 10³ MPa)	66.9	41.3	19.5	16.1	30.0	4.4
Permeability (× 10⁻⁹ m/s)	0.3	40.9	125.4	160.5	1.4	13.9

Rainwater is generally weakly acidic and further acids may be taken into solution from organic or mineral matter. The degree of aggressiveness of water to limestone can be assessed on the basis of the relationship between the dissolved carbonate content, the pH value and the temperature of the water. At any given pH value, the cooler the water the more aggressive it is. If solution continues its rate slackens and it eventually ceases when saturation is reached. Hence solution is greatest when the bicarbonate saturation is low. This occurs when water is circulating so that fresh supplies with low lime saturation are continually made available. Non-saline water can dissolve up to 400 ppm of calcium carbonate.

Limestones are commonly transected by joints. These have generally been subjected to varying degrees of solutioning so that some may gape. Sinkholes may develop where joints intersect and these may lead to subterranean galleries and caverns. The latter are characteristic of thick, massive limestones. Sometimes solutioning produces a highly irregular, pinnacled surface on limestone pave-ments. The size, form, abundance, and downward extent of the aforementioned features depends upon the geological structure and the presence of interbedded impervious layers. Individual cavities may be open; they may be partially or completely filled with clay, silt, sand or gravel mixtures, or they may be water-filled conduits. Solution cavities present numerous problems in the construction of large foundations such as for dams, among which bearing strength and water-tightness are paramount.

Few sites are so bad that it is impossible to construct safe and successful structures upon them, but the cost of the necessary remedial treatment may be prohibitive. Dam sites should be abandoned where the cavities are large and numerous and extend to considerable depths.

The primary effect of solution in limestone is to enlarge the pores which enhances water circulation, encouraging further solution. This brings about an increase in stress within the remaining rock framework which reduces the strength of the rock mass and leads to increasing stress corrosion. On loading, the volume of the voids is reduced by fracture of the weakened cement between the particles and by the reorientation of the intact aggregations of rock that become separated by loss of bonding. In such situations settlement may occur, most of which takes place within a few days of the application of load. Sowers[13] related a case of settlement of two and three-storey reinforced concrete buildings founded on soft porous oolitic limestone in Florida which had been subjected to solutioning. The limestone had been leached and was highly porous with little induration remaining 2.1 m below the rock surface. The building was surrounded by a crack in the more intact surface crust and settlement exceeding 100 mm was recorded. Standard penetration resistances obtained from the weakened oolite ranged between 1 and 2 blows per 0.3 m.

Differential settlement under loading has occurred in limestone which at ground level appeared competent, but immediately beneath the surface consisted of long, narrow pinnacles of rock separated by solution channels occupied by broken limestone rubble in a clay matrix.

Rapid subsidence can take place due to the collapse of holes and cavities within limestone which has been subjected to prolonged solution, this occurring when the roof rocks are no longer thick enough to support themselves. It must be emphasised, however, that the solution of limestone is a very slow process; contemporary solution is therefore very rarely the cause of collapse. For instance, Kennard and Knill[14] quoted mean rates of surface lowering of limestone areas in the British Isles which ranged from 0.041 to 0.099 mm annually. They also quoted experiments carried out in flowing non-saline water which produced a solution rate of approximately 1 mm per year.

Nevertheless, solution may be accelerated by man-made changes in the ground water conditions or by a change in the character of the surface water that drains into limestone. As an example, solution rates varying between 1.5 and 3.1 mm per year were recorded at the Hales Bar reservoir on the river Tennessee (see Frink[15]). North[16] described collapses of cavities in the Lias Limestone in Bridgend. These collapses were caused by ancient cavities being enlarged by excessive water discharging through them. In one case water was escaping from a water main, in another leakage was occurring from storm drains. North noted that in one case where a collapsed cavity was filled it collapsed again the following year. The reason given was that the supporting fill was washed by flowing water from the cavity into a fissure. The obvious conclusion is that where possible such troublesome cavities should be lined prior to treatment with fill.

According to Sowers[13] ravelling failures are the most widespread and probably the most dangerous of all the subsidence phenomena associated with limestones. Ravelling occurs when solution-enlarged openings extend upward to a rock surface overlain by soil. The openings should be interconnected and lead into channels through which the soil can be eroded by ground water flow.

Initially the soil arches over the openings but as they are enlarged a stage is reached when the soil above the roof can no longer support itself and so it collapses. A number of conditions accelerate the development of cavities in the soil and initiate collapse. Rapid changes in moisture content produce aggravated slabbing or roofing in clays and flow in cohesionless sands. Lowering the water table increases the downward seepage gradient and accelerates downward erosion; reduces capillary attraction in sand and increases the ability to flow through narrow openings; and gives rise to shrinkage cracks in highly plastic clays which weakens the mass in dry weather and produces concentrated seepage during rains. Increased infiltration often initiates failure, particularly when it follows a period when the water table has been lowered.

Solution features are also developed in dolostone. Jennings *et al.*[17] described the collapse of a three-storeyed crusher plant into a sinkhole at Westdriefontein. They noted that some of the sinkholes in the Transvaal Dolomite were 100 m in diameter and 40 m in depth. This particular collapse took place suddenly and the authors suggested that it was influenced by the lowering of the water table, by up to 120 m, due to pumping from mines. This meant that fissures in the roofs of caverns in the dolostone were enlarged by solutioning due to perculating water. Avens (i.e. chimneys) were developed in the roofs and they spread into the overlying mantle of soil. Hence the arches above individual caverns were seriously weakened.

With the exception of certain horizons in the Lower Chalk of south east England which possesses an appreciable mud content, the Chalk is a remarkably pure, soft limestone usually containing over 95% calcium carbonate. Generally chalk can be divided into coarse and fine fractions. The coarse fraction, which may constitute 20 to 30%, falls within the 10 to 100 micron range. This contains material derived from the mechanical breakdown of large shelled organisms and, to a lesser extent, from foraminifera. The fine fraction, which takes the form of calcite particles about one micron or less in size, is almost entirely composed of coccoliths and may form up to, and sometimes over, 80% of certain horizons.

Bell[18] found a notable range in the dry density of chalk, which substantiated the work of Higginbottom[19]. For example, low values have been recorded from the Upper Chalk of Kent (1.35 to 1.64 Mg/m^3) whilst those from the Middle Chalk of Norfolk and the Lower Chalk of Yorkshire frequently exceed 2.0 Mg/m^3. In other words the density of the Chalk tends to increase with depth, presumably due to increasing overburden pressures, the void ratio decreasing. The porosity of chalk tends to range between 30 and 50%.

Carter and Mallard[20] found that chalk compressed elastically up to a critical pressure, the apparent preconsolidation pressure. Marked breakdown and substantial consolidation occurs at higher pressures. The apparent preconsolidation pressure is influenced by consolidation, cementation and possibly creep. They obtained coefficients of consolidation (c_v) and compressibility (m_v) similar to those found by Meigh and Early[21] and Wakeling[22]; $c_v = 1135$ m^2/yr, $m_v = 0.019$ m^2/MN. The unconfined strength of the Chalk ranges from moderately weak (much of the Upper Chalk) to moderately strong (much of the Lower Chalk of Yorkshire and the Middle Chalk of Norfolk) according to the strength classification recommended by the Geological Society[23]. However, the unconfined compressive strength of chalk undergoes a marked reduction when it is saturated. For example, according to Bell[18], some samples of Upper Chalk from

Samples from the Lower and Middle Chalk may show a reduction in strength
Kent suffer a dramatic loss on saturation amounting to approximately 70%.
averaging some 50%. The pore water also has a critical influence on the triaxial
strengths of chalk and obviously its modulus of deformation (see Meigh and
Early[21]).

Bell[18] noted that the mode of failure of the Upper Chalk when tested in
triaxial conditions was influenced by the confining pressure. Diagonal shear
failure occurred at the lower confining pressures but at 4.9 MPa confining
pressure and above plastic deformation took place, giving rise to barrel shaped
failures in which numerous small, inclined shear planes were developed. Meigh
and Early[21] suggested that at high cell pressures the sample underwent disag-
gregation, which they demonstrated by wetting and drying failed samples, this
causing a collapse of their structure.

Chalk is a non-elastic rock (E_i = less than 5×10^4 MPa), and the Upper
Chalk from Kent is particularly deformable, a typical value of Young's modulus
being 5×10^3 MPa. In fact the Upper Chalk exhibits elastic-plastic deformation, with perhaps incipient creep, prior to failure. The deformation
properties of chalk in the field depend upon its hardness, and the spacing, tightness and orientation of its discontinuities. These values are also influenced by
the amount of weathering it has undergone. Using these factors, Ward *et al*[24]
classified the Middle Chalk at Mundford, Norfolk, into five grades, and showed
that the value of Young's modulus varies with grade (Table 8.4). They pointed
out that grades IV and V were largely the result of weathering and were therefore independent of lithology whereas grades I and II were completely
unweathered so that the difference between them was governed by their lithological character. Accordingly grades V, IV and III occur in succession from the
surface down whilst grade I may overlie grade II or vice versa.

Burland and Lord[27] observed that both the full-scale tank test and plate load
tests at Mundford indicated that at low applied pressures even grade IV chalk
behaves elastically. At higher pressures chalk exhibits yielding behaviour. They
pointed out that for stresses up to 1.0 MPa the plate load tests showed that
grades IV and V exhibit significant creep, and in the long term creep deflections
may be considerably larger than immediate deflections, creep in grade III is
smaller and terminates more rapidly whilst grades II and I undergo negligible
creep.

Burland *et al*[28] found that settlements of a five-storey building founded in
soft low grade chalk at Reading were very small. Their findings agreed favourably
with those previously obtained at Mundford, as did the results of an investigation
carried out at Basingstoke by Kee *et al*[29].

As in limestone, the discontinuities are the fundamental factors governing the
mass permeability of chalk (see Ineson[30]). Chalk is also subject to solutioning
along discontinuities. However, subterranean solution features tend not to
develop in chalk since it is usually softer than limestone and so collapses as
solution occurs. Nevertheless solution pipes and swallow holes are present in the
Chalk, being commonly found near the contact of the Chalk with the overlying
Tertiary and drift deposits. West and Dumbleton[31] suggested that high concentrations of water, for example, run-off from roads, can lead to the re-activation
of swallow holes and the formation of small pipes within a few years. They also
recorded that new swallow holes often appear at the surface without warning
after a period of heavy rain or following the passage of plant across a site. More-

Table 8.4 CORRELATION BETWEEN GRADES AND THE MECHANICAL PROPERTIES OF MIDDLE CHALK AT MUNDFORD (after Ward et al.*)

Grade†	Description	Approximate range of E (MPa)‡	Bearing pressure causing yield (kPa)	Creep properties	SPT N Value† (after Wakeling[25])	Rock mass factor (after Burland and Lord[27])
V	Structureless melange. Unweathered and partly weathered angular chalk blocks and fragments set in a matrix of deeply weathered remoulded chalk. Bedding and jointing are absent	Below 500	Below 200	Exhibits significant creep	Below 15	0.1
IV	Friable to rubbly chalk. Unweathered or partially weathered chalk with bedding and jointing present. Joints and small fractures closely spaced, ranging from 10–60 mm apart	500–1000	200–400	Exhibits significant creep	15–20	0.1 to 0.2
III	Rubbly to blocky chalk. Unweathered medium to hard chalk with joints 60–200 mm apart. Joints open up to 8 mm, sometimes with secondary staining and fragmentary infillings	1000–2000	400–600	For pressures not exceeding 400 kPa creep is small and terminates in a few months	20–25	0.2 to 0.4
II	Medium hard chalk with widely spaced, closed joints. Joints more than 200 mm apart. Fractures irregularly when excavated, does not break along joints. Unweathered	2000–5000	Over 1000	Negligible creep for pressure of at least 400 kPa	25–35	0.6 to 0.8
I	Hard, brittle chalk with widely spaced, closed joints. Unweathered	Over 5000	Over 1000	Negligible creep for pressure of at least 400 kPa	Over 35	Over 0.8

*Ward et al.[24] emphasized that their classification was specifically developed for the site at Mundford and hence its application elsewhere should be made with caution

†The correlation between SPT N value and grade may be different in the Upper Chalk (see Dennehy[26])

‡See also 1. Abbiss, C.P. 'A comparison of the stiffness of the Chalk at Mundford from a seismic survey and large-scale tank test.' *Geotechnique*, **29**, 461–8, (1979).

2. Grainger, P., McCann, D.M. and Gallois, R.W. 'The application of seismic refraction to the study of fracturing in the Middle Chalk at Mundford, Norfolk. *Geotechnique*, **23**, 219–232 (1973).

over they found that voids can gradually migrate upwards through chalk due to material collapsing. Lowering of the chalk surface beneath overlying deposits due to solution can occur, disturbing the latter deposits and lowering their degree of packing. Hence the chalk surface may be extremely irregular in places.

Chalk during cold weather may suffer frost heave (see Lewis and Croney[32]), ice lenses up to 25 mm in thickness being developed along bedding planes. Higginbottom[19] suggested that a probable volume increase of some 20 to 30% of the original thickness of the ground may ultimately result.

Solifluction deposits known as head or coombe deposits, and which were formed under periglacial conditions during Pleistocene times, are commonly found along valley bottoms carved in the Chalk of southern England. Head is a poorly strati-fied, poorly sorted deposit of angular chalk fragments, frequently set in a matrix of remoulded, pasty, fine chalk detritus. Coombe deposits are to a varying degree cemented with secondary carbonate. Frost shattering also took place during Pleistocene times and its effects in the Chalk of southern England may extend to depths of several metres.

8.4 EVAPORITIC ROCKS

Representative relative densities and dry densities of gypsum, anhydrite, rock salt and potash are given in Table 8.5, as are the porosity values. Anhydrite according to the classification of unconfined compressive strength (Geological Society[23]) is a strong rock, gypsum and potash are moderately strong, whilst rock salt is moderately weak (Table 8.5). Values of Young's modulus are also given in Table 8.5, from which it can be ascertained that gypsum and anhydrite have high values of modulus ratio whilst potash and rock salt have medium values. Evaporitic rocks exhibit varying degrees of plastic deformation prior to failure, for example, in rock salt the yield strength may be as little as one tenth the ultimate compressive strength, whereas anhydrite undergoes comparatively

Table 8.5 SOME PHYSICAL PROPERTIES OF EVAPORITIC ROCKS

	Gypsum (Sherburn in Elmet)	Anhydrite (Sandwith)	Rock salt (Winsford)	Potash (Loftus)
Relative density	2.36	2.93	2.2	2.05
Dry density (Mg/m³)	2.19	2.82	2.09	1.98
Porosity (%)	4.6	2.9	4.8	5.1
Unconfined compressive strength (MPa)	27.5	97.5	11.7	25.8
Point load strength (MPa)	2.1	3.7	0.3	0.6
Schleroscope hardness	27	38	12	9
Schmidt hardness	25	40	8	11
Young's modulus (× 10³ MPa)	24.8	63.9	3.8	7.9
Permeability (× 10⁻¹⁰ m/s)	6.2	0.3	–	–

little plastic deformation. Creep may account for anything between 20 and 60% of the strain at failure when these evaporitic rocks are subjected to incremental creep tests. Rock salt is most prone to creep. However, Justo and Zapico[33] recorded that the amount of settlement which occurred when gypsum was subjected to plate load testing, the maximum load being 1.2 MPa, was negligible.

Gypsum is more readily soluble than limestone, for example, 2100 ppm can be dissolved in non-saline waters as compared with 400 ppm. Sinkholes and caverns can therefore develop in thick beds of gypsum (see Eck and Redfield[34]) more rapidly than they can in limestone. Indeed in the United States they have been known to form within a few years where beds of gypsum are located beneath dams. Extensive surface cracking and subsidence has occurred in parts of Oklahoma and New Mexico due to the collapse of cavernous gypsum (see Brune[35]). The problem is accentuated by the fact that gypsum is weaker than limestone and therefore collapses more readily.

Kendal and Wroot[36] quoted vivid accounts of subsidences which occurred in the Ripon area in the eighteenth and nineteenth centuries due to the solution of gypsum. They wrote that wherever beds of gypsum approach the surface craters have been formed by the collapse of overlying rocks into areas from which gypsum has been removed by solution. However, where gypsum is effectively sealed from the ingress of water by overlying impermeable strata such as marl, solutioning does not occur (see Redfield[37]). The solution of gypsum gives rise to sulphate bearing ground waters which mean that normal concrete may suffer accordingly.

The solution rate of gypsum or anhydrite is controlled principally by the surface area in contact with water and the flow velocity of water associated with a unit area of the material. Hence the amount of fissuring in a rock mass, and whether it is enclosed by permeable or impermeable beds, is important. Solution also depends on the sub-saturation concentration of calcium sulphate in solution. According to James and Lupton[38], the concentration dependence for gypsum is linear while that for anhydrite is a square law. The salinity of the water also is influential. For example, the rates of solution of gypsum and anhydrite are increased by the presence of sodium chloride, carbonate and carbon dioxide in solution. It is therefore important to know the chemical composition of the ground water.

Gypsum is usually less dangerous than anhydrite because it tends to dissolve in a steady manner forming caverns or causing progressive settlements. For instance, if small fissures occur at less than 1 m intervals, solution usually occurs by removal of gypsum as a front moving 'downstream' at less than 0.01 m/year. However, James and Lupton showed that if:

(1) the rock temperature were 10°C,
(2) the water involved contained no dissolved salts,
(3) and a hydraulic gradient of 0.2 were imposed,

then a fissure, 0.2 mm wide and 100 m long, in massive gypsum, would in 100 years have widened by solution so that a block 1 m^3 in size could be accommodated in the entrance to the fissure. In other words a cavern would be formed. If the initial width of the fissure exceeds 0.6 mm, large caverns would form and a runaway situation could develop in a very short time. In long fissures the hydraulic gradient is low and the rate of flow is reduced so that solutions become saturated and little or no material is removed. Indeed James and Lupton implied that a flow rate of 10^{-3} m/s was rather critical in that if it were exceeded, extensive solution of gypsum could take place. Solution of massive gypsum is not likely to give rise to an accelerating deterioration in a foundation if precautions such as grouting are taken to keep seepage velocities low. On the

other hand gypsum is most hazardous when it forms the cement in a conglo-
merate since solution of small amounts can reduce the strength of the rock to a
very low value.

Hawkins[39] noted the presence of voids in rocks of Keuper age in the area
around the Severn estuary. These voids were formed as a result of gypsum being
removed in solution. He also noted the effects of solutioning in dolomitic rocks
which contain gypsum. This action may lead to an enrichment in calcium, as
magnesium sulphate is lost in a solution.

Massive anhydrite can be dissolved to produce uncontrollable runaway
situations in which seepage flow rates increase in a rapidly accelerating manner.
Even small fissures in massive anhydrite can prove dangerous. If anhydrite is
taken in the above example not only is a cavern formed but the fissure is
enlarged as a long tapering section. Within about 13 years the flow rate increases
to a runaway situation. However, if the fissure is 0.1 mm in width then the
solution becomes supersaturated with calcium sulphate and gypsum is
precipitated. This seals the outlet from the fissure and from that moment any
anhydrite in contact with the water is hydrated to form gypsum. Accordingly
0.1 mm width seems to be a critical fissure size in anhydrite. Such conversion is
characteristic of extensive deposits of permeable granular anhydrite. Anhydrite
is less likely to undergo catastrophic solution in a fragmented or particulate
form than gypsum.

Uplift is a problem associated with anhydrite. This takes place when anhydrite is hydrated to form gypsum; in so doing there is a volume increase of
between 30 and 58% which exerts pressures that have been variously estimated
between 2 and 69 MPa. It is thought that no great length of time is required
to bring about such hydration. When it occurs at shallow depths it causes expansion but the process is gradual and is usually accompanied by the removal of
gypsum in solution. At greater depths anhydrite is effectively confined during
the process. This results in a gradual build-up of pressure and the stress is finally
liberated as an explosive force. According to Brune[34] such uplifts in the United
States have taken place beneath reservoirs, these bodies of water providing a
constant supply for the hydration process, percolation taking place via cracks
and fissures. Examples are known of the ground surface being elevated by about
6 m. The rapid, explosive movement causes strata to fold, buckle and shear
which further facilitates access of water into the ground.

Salt is even more soluble than gypsum and the evidence of slumping,
brecciation and collapse structures in rocks which overlie saliferous strata bear
witness to the fact that salt has gone into solution in past geological times. It is
generally believed, however, that in areas underlain by saliferous beds
measurable surface subsidence is unlikely to occur except where salt is being
extracted. Perhaps this is because equilibrium has been attained between the
supply of unsaturated ground water and the salt available for solution. Exceptionally, cases have been recorded of rapid subsidence, e.g. the 'Meade salt sink'
in Kansas was explained by Johnson[40] as due to solution of deep salt beds. This
area of water, about 60 m in diameter, occurred as a result of subsidence in
March 1879. At the same time, 64 km to the south west, the railway station at
Rosel and several buildings disappeared due to the sudden appearance of a sinkhole.

The occurrence of salt springs in Cheshire has been quoted as evidence of
natural solution and it is thought that the Cheshire meres were formed as a result

shales and mudstones, one grading into the other. Shales, however, are characterised by their lamination. Shale differs from residual clay in grain size distribution, the average shale consisting of about two thirds silty material and one
third clay fraction. There are, of course, exceptions. Non-silty siliceous shales
are usually diatom rich or derived from volcanic ash. Bauxitic shales, black
pyritous shales and sideritic shales are also usually fine grained. Shales may also
contain appreciable quantities of lime or gypsum, indeed calcareous shales frequently grade into shaley limestone. Carbonaceous shales are those which
accumulated slowly under anaerobic conditions and are rich in sulphur
compounds.

Quartz usually accounts for approximately one third of a normal shale, clay
minerals, including micas and chlorite, for another third and other minerals such
as feldspar, calcite, dolomite, pyrite, hematite and limonite, together with some
carbonaceous matter, make up the remainder. The mineral content of shales
influences their geotechnical properties, the most important factor in this
respect being the quartz-clay minerals ratio. For example, the liquid limit of clay
shales increases with increasing clay mineral content, the amount of montmorillonite, if present, being especially important. Mineralogy also affects the activity
of an argillaceous material, again this increases with clay mineral content,
particularly with increasing content of montmorillonite. Activity influences the
slaking characteristics of a shale.

Consolidation with concomitant recrystallisation on the one hand and the
parallel orientation of platey minerals, notably micas, on the other give rise to
the fissility of shales. An increasing content of siliceous or calcareous material
gives a less fissile shale whilst carbonaceous shales are exceptionally fissile.
Moderate weathering increases the fissility of shale by partially removing the
cementing agents along the laminations or by expansion due to the hydration of
clay particles. Intense weathering produces a soft clay-like soil. Ingram[43]
recognised massive, flaggy and flakey varieties of shale according to their degree
of fissility. Some shales exhibit all degrees of fissility in the same bed.

Lamination effects an important control on the breakdown of shales (see
Taylor and Spears[48]). Other controls on the breakdown of shaley materials
include air breakage and dispersal of colloidal material (see Chapter 7 and
Badger *et al*[49]). A feature of the breakdown of shales and mudstones is their
disintegration to produce silty clays (see Grice[50]).

Shale is occasionally a notoriously difficult, and often an undesirable material
to work in. Certainly there have been many failures of structures and slopes in
shales. Nevertheless many shales have proved satisfactory as foundation rocks.
Hence it can be concluded that shales vary widely in their engineering behaviour
and that it is therefore necessary to determine the problematic types. This
variation in engineering behaviour to a large extent depends upon their degree of
compaction and cementation, indeed Mead[44] divided shales into compaction and
cementation types in his classification (Table 8.7).

The cemented shales are invariably stronger and more durable. Marine shales
may be impregnated with carbonate cement which makes them appreciably
stronger and indeed they may grade into impure limestone. Carbonaceous shales
contain a significant proportion of organic matter and are therefore softer.

The natural moisture content of shales varies from less than 5%, increasing to
as high as 35% for some clay shales. When the natural moisture content of shales
exceeds 20% they frequently are suspect as they tend to develop potentially high

of the subsidence which thereby occurred, but such subsidence usually is a ve slow process operating over large areas. Localised solution is presumed develop collecting channels along the upper surface of salt beds. The incor petence of the overlying Keuper Marl inhibits the formation of large voids, roc collapse no doubt taking place at more or less the same time as salt removal. Th action eventually gives rise to linear subsidence depressions or brine runs i ground level.

Classic examples of subsidence due to salt working have occurred in Cheshir where salt has been extracted for over 300 years. There the salt occurs at tw principal horizons. Natural brine pumping, which still continues, has produce surface troughs of various sizes and shapes, depending on the amount o pumping and the competency of the overlying rocks. As the brine runs extend s the subsidence and existing depressions are enlarged and deepened. It is usuall impossible to predict the extent and amount of future subsidence since th width and height of the area undergoing solution is almost always unknown (se Bell[41]).

8.5 SILTSTONES

Siltstones may be massive or laminated, the individual laminae being picked out by mica and/or carbonaceous material. Micro-cross bedding is frequently developed and in some siltstones the laminations may be convoluted. Siltstones have a high quartz content with a predominantly siliceous cement. They there-fore tend to be hard, tough rocks (Table 8.6). Frequently siltstones are inter-bedded with shales or fine grained sandstones, the siltstones occurring as thin

Table 8.6 ENGINEERING PROPERTIES OF SOME COAL MEASURES ROCKS

	Mudstone	Siltstone	Shale	Barnsley Hards Coal	Deep Duffryn Coal
Relative density	2.69	2.67	2.71	1.5	1.2
Dry density (Mg/m³)	2.32	2.43	2.35	–	–
Dry unconfined compressive strength (MPa)	45.5	83.1	20.2	54.0	18.1
Saturated unconfined com- pressive strength (MPa)	21.3	64.8	–	–	–
Point load strength (MPa)	3.8	6.2	–	4.1	0.9
Scleroscope hardness	32	49	–	–	–
Schmidt hardness	27	39	–	–	–
Young's modulus (× 10³ MPa)	25	45	5.2	26.5	–

ribs. Like sandstones, their disintegration is governed by their fracture pattern. After several months of weathering debris in excess of cobble size may be pro-duced. Subsequent degradation down to component grain size takes place at a very slow rate.

8.6 SHALES

Shales are the most abundant sediments accounting for approximately half the stratigraphical column (see Kuenen[42]). There is no sharp distinction between

pore pressures. Usually the moisture content in the weathered zone is higher than in the unweathered shale beneath.

Dependent upon the relative humidity, many shales slake almost immediately when exposed to air (see Kennard *et al*[45]). Desiccation of shale, following exposure, leads to the creation of negative pore pressures and consequent tensile failure of the weak intercrystalline bonds. This in turn leads to the production of shale particles of coarse sand, fine gravel size. Alternate wetting and drying causes a rapid breakdown of compaction shales.

Low grade compaction shales undergo complete disintegration after several cycles of drying and wetting, whilst well cemented shales are resistant. Indeed De Graft-Johnson *et al*[46] found that the compacted variety of Accra Shale could be distinguished from the cemented variety by wetting and drying tests. The compacted shales generally crumbled to fine material after 2 or 3 cycles whilst the cemented samples withstood 6 cycles, none of the samples exceeding a loss of 8%.

Morgenstern and Eigenbron[47] used a water absorption test to assess the amount of slaking undergone by argillaceous material. This test measures the increase of water content in relation to the number of drying and wetting cycles undergone. They found that the maximum slaking water content increased linearly with increasing liquid limit and that during slaking all materials eventually reached a final water content equal to their liquid limit. Materials with medium to high liquid limits, in particular, exhibited very substantial volume changes during each wetting stage, which caused large differential strains,

Table 8.7 CLASSIFICATION OF SHALE (After Mead[44])

SHALE		
COMPACTION SHALE	**CEMENTED SHALE**	
CLAY SHALE	50% or more clay particles which may or may not be true minerals	CALCAREOUS SHALE — 20% to 35% CaCo$_3$ (Marls and shaly chalk 35% to 65% CaCo$_3$)
SILTY SHALE	25% to 45% silt sized particles. Silt may be in thin layers between clayey shale bands	SILICEOUS SHALE — 70%–85% amorphous silica often highly siliceous volcanic ash (quartzose shale — detrital quartz)
SANDY SHALE	25% to 45% sand sized particles. Sand may be in thin layers between clayey shale bands	FERRUGINOUS SHALE — (25%–35% Fe$_2$O$_3$) (Potassic shale — 5%–10% potash)
BLACK SHALE	Organic rich, splits into thin semi-flexible sheets	CARBONACEOUS SHALE (Oil shale, Bone coal) — Carbonaceous matter (3% to 15%) tends to bond constituents together and imparts a certain degree of toughness
		CLAY BONDED SHALE — Welded by recrystallization of clay minerals, or by other diagenetic bonds

Table 8.8 ENGINEERING EVALUATION OF SHALES (After Underwood[52])

Laboratory tests and in situ observations	Physical properties — Average range of values		Probable in situ behaviour						
	Unfavourable	Favourable	High pore pressure	Low bearing capacity	Tendency to rebound	Slope stability problems	Rapid sinking	Rapid erosion	Tunnel support problems
Compressive strength (kPa)	350–2070	2070–3500	✓	✓					✓✓
Modulus of elasticity (MPa)	140–1400	1400–14000			✓✓	✓✓			✓✓
Cohesive strength (kPa)	35–700	700–>10500			✓✓	✓✓			✓
Angle of internal friction, degrees	10–20	20–65			✓✓	✓✓			
Dry density Mg/m³	1.12–1.78	1.78–2.56	✓		✓			✓(?)	
Potential swell (%)	3–15	1–3						✓✓	
Natural moisture content (%)	20–35	5–15	✓			✓✓			
Coefficient of permeability m/s	10⁻⁷–10⁻¹²	>10⁻⁷	✓✓				✓		
Predominant clay minerals	Montmorillonite or illite	Kaolinite and chlorite	✓✓						
Activity ratio	0.75–>20	0.35–0.75				✓			
Wetting and drying cycles	Reduces to grain sizes	Reduces to flakes					✓	✓	
Spacing of rock defects	Closely spaced	Widely spaced		✓		✓✓			
Orientation of rock defects	Adversely oriented	Favourable oriented		✓		✓✓		✓(?)	
State of stress	> Existing overburden load	≅ Overburden load			✓	✓✓			✓

Note: According to S. Irmay (Israel Journal of Technology, Vol 6, No. 4, pp 165–172, 1968), the maximum possible $\phi = 47.4°$. The ticks relate to the unfavourable range of values.

resulting in complete destruction of the original structure. Thus materials characterised by high liquid limits are more severely weakened during slaking than materials with low liquid limits.

The primary problem attributable to slaking of shale during construction is that upon exposure it becomes coated with mud when wetted, which prevents the development of bond between concrete and rock. This can be prevented by coating the surface with a protective material, or by pouring a protective concrete cover immediately after exposure. Slaking of shales after construction causes ravelling and spalling of cut slopes and is sometimes the cause of the undermining and collapse of more competent beds, but this is rarely a serious problem. Mudstones tend to break down along irregular fracture patterns, which when well developed, can mean that these rocks disintegrate within one or two cycles of wetting and drying.

The swelling properties of certain clay shales have proved extremely detrimental to the integrity of many civil engineering structures. Swelling is attributable to the absorption of free water by certain clay minerals, notably montmorillonite, in the clay fraction of a shale. Casagrande[51] noted that highly fissured overconsolidated shales have greater swelling tendencies than poorly fissured clay shales, the fissures providing access for water.

The degree of packing, and hence the porosity, void ratio and density of a shale, depend on its mineral composition and grain size distribution, its mode of sedimentation, its subsequent depth of burial and tectonic history, and the effects of diagenesis. The porosity of shale may range from slightly under 5 to just over 50%, with natural moisture contents of 3 to 35%. Argillaceous materials are capable of undergoing appreciable suction before pore water is removed, drainage commencing when the necessary air-entry suction is achieved (about pF 2). Under increasing suction pressure the incoming air drives out water from a shale and some shrinkage takes place in the fabric before air can offer support. Generally as the natural moisture content and liquid limit increase so the effectiveness of suction declines.

It would appear that the strength of compacted shales decreases exponentially with increasing void ratio and moisture content. However, in cemented shales the amount and strength of the cementing material are the important factors influencing intact strength. The value of shearing resistance of a shale chosen for design purposes will lie somewhere between the peak and residual strengths. In weak compaction shales cohesion may be lower than 15 kPa and the angle of friction as low as 5°. By contrast, Underwood[52] quoted values of cohesion and angle of friction of 750 kPa and 56° respectively for dolomitic shales of Ordovician age, and 8 to 23 MPa and 45° to 64° respectively for calcareous and quartzose shales from the Cambrian period. Generally shales with a cohesion less than 20 kPa and an apparent angle of friction of less than 20° are likely to present problems. The elastic moduli of compaction shales range from less than 140 to 1400 MPa, whilst well cemented shales have elastic moduli in excess of 14000 MPa (see Table 8.8). Unconfined compressive strength tests on Accra Shales carried out by De Graft-Johnson *et al*[46] indicated that the samples usually failed at strains between 1.5 and 3.5%. The compressive strengths varied from 200 kPa to 20 MPa, with the values of the modulus of elasticity ranging from 6.5 to 1400 MPa. Those samples which exhibited the high strengths were generally cemented types.

Morgenstern and Eigenbrod[47] carried out a series of compression softening tests on argillaceous materials. The rate of softening when immersed in water depends largely upon the degree of induration of the material involved, poorly indurated materials softening very quickly and they may undergo a loss of up to 90% of their original strength within a few hours. Mudstones at their natural water contents remain intact when immersed in water, however, they swell slowly, hence decreasing in bulk density and strength. This time-dependent loss in strength is a very significant engineering property of mudstones. A good correlation exists between initial compressive strength and the amount of strength loss during softening.

According to Zaruba and Bukovansky[53] the mechanical properties of the Ordovician shales in the Prague area are controlled by their lithological composition, degree of weathering and amount of tectonic disturbance. They quoted results of numerous loading tests which indicated that the moduli of elasticity were influenced more by the degree of weathering than by the lithological composition of the rock. However, when tests were performed in galleries on unweathered shales, the material, which was predominantly pelitic, gave lower values of the moduli of elasticity than that which was predominantly quartzose. The higher the degree of fissility possessed by a shale the greater the anisotropy with regard to strength, deformation and permeability. For instance, the influence of fissility on Young's modulus can be illustrated by two values quoted by Chappell[54], 6000 and 7250 MPa, for cemented shale tested parallel and normal to the lamination respectively.

Zaruba and Bukovansky[53] found that the values of Young's modulus were up to five times greater when they tested shale normal as opposed to parallel to the direction of lamination.

According to Burwell[55] well-cemented shales, under structurally sound conditions, present few problems for large structures such as dams, though their strength limitations and elastic properties may be factors of importance in the design of concrete dams of appreciable height. They, however, have lower moduli of elasticity and generally lower shear values than concrete and therefore in general are unsatisfactory foundation materials for arch dams.

The problem of settlement in shales generally resolves itself into one of reducing the unit bearing load by widening the base of structures or using spread footings. In some cases appreciable differential settlements are provided for by designing articulated structures capable of taking differential movements of individual sections without damage to the structure. Severe settlements may take place in low-grade compaction shales. However, compaction shales contain fewer open joints or fissures which can be compressed beneath heavy structures, than do cemented shales. Where concrete structures are to be founded on shale and it is suspected that the structural load will lead to closure of defects in the rock, in situ tests should be conducted to determine the elastic modulus of the foundation material.

Uplift frequently occurs in excavations in shales and is attributable to swelling and heave. Rebound on unloading of shales during excavation is attributed to heave due to the release of stored strain energy. The conserved strain energy tends to be released more slowly than in harder rocks. Shale relaxes towards a newly excavated face and sometimes this occurs as offsets at weaker seams in the shale. The greatest amount of rebound occurs in heavily over-consolidated compaction shales; for example, at Garrison Dam, North Dakota, just over 0.9 m of rebound was measured in the deepest excavation in the Fort Union clay shales. What is more, high horizontal residual stresses caused saw cuts

75 mm in width to close within 24 h (see Smith and Redlinger[56]). Some 280 mm of rebound was recorded in the underground workings in the Pierre Shale at Oake Dam, South Dakota (see Underwood *et al*[57]). Moreover differential rebound occurred in the stilling basin due to the presence of a fault. Differential rebound movements require special design provision.

The stability of slopes in excavations can be a major problem in shale both during and after construction. This problem becomes particularly acute in dipping formations and in formations containing expansive clay minerals.

Sulphur compounds are frequently present in shales, clays, mudstones and marls. An expansion in volume large enough to cause structural damage can occur when sulphide minerals such as pyrite and marcasite suffer oxidation to give anhydrous and hydrous sulphates.

According to Fasiska, Wagenblast and Dougherty[58] the pyrite structure may be regarded as a stacking of almost close packed hexagonal sheets of sulphide ions with iron ions in the interstices between the sulphide layers. The packing density is related to the radius of the sulphide ion which is 1.85 Å (volume = 26.1 Å). In the sulphate structure each atom of sulphur is surrounded by four atoms of oxygen in tetrahedral coordination. The packing density is related to the radius of the sulphate ion which is 2.8 Å, giving a volume of 92.4 Å. This represents an increase in volume per packing unit of approximately 35%. Hydration involves a further increase in volume. In fact such a reaction is electrolytic, that is, water is required and the sulphate ion exists in solution. Any cation in the system may cause the precipitation of sulphate crystals. If calcium carbonate is present, gypsum may be formed, which may give rise to an eightfold increase in volume over the original sulphide, exerting pressures of up to about 0.5 MPa. This leads to further disruption and weakening of the rocks involved.

Penner *et al*[59] quoted a case of heave in a black shale of Ordovician age in Ottawa which caused displacement of the basement floor of a three storey building. The maximum movement totalled some 107 mm, the heave rate being almost 2 mm per month. When examined the shale in the heaved zone was found to have been altered to a depth of between 0.7 and 1 m. Beneath, the unaltered shale contained numerous veins of pyrite, indeed the sulphur content was as high as 1.6%. The heave was attributable to the alteration of pyrite to gypsum and jarosite, these forming in the fissures and between the laminae of the shales in the altered zone. Measurements of the pH gave values ranging from 2.8 to 4.4, and it was therefore concluded that the alteration of the shales was the result of biochemical weathering brought about by autotrophic bacteria. The heaving was therefore arrested by creating conditions unfavourable for bacterial growth. This was done by neutralising the altered zone by introducing a potassium hydroxide solution into the examination pits. The water table in the altered zone was also kept artificially high so that the acids would be diffused and washed away, and to reduce air entry.

Similarly sulphuric acid and sulphate are produced when gypsum is subjected to weathering. Aqueous solutions of sulphate and sulphuric acid react with tricalcium aluminate in Portland cement to form calcium sulpho-aluminate or ettringite. This reaction is accompanied by expansion. The rate of attack is very much influenced by the permeability of the concrete or mortar and the position of the water table. For example, sulphates can only continue to reach cement by movement of their solutions in water. Thus if a structure is permanently above the water table it is unlikely to be attacked.

By contrast below the water table movement of water may replenish the sulphates removed by reaction with cement, thereby continuing the reaction. Concrete with a low permeability is essential to resist sulphate attack, hence it should be fully compacted. Sulphate resistant cements, that is, those in which the tricalcium aluminate is low can also be used for this purpose (Table 8.8). Foundations used to be made larger than necessary to counteract sulphate attack but it is now more economical to protect them by impermeable membranes or bituminous coatings.

When a load is applied to an essentially saturated shale foundation the void ratio in the shale decreases and the pore water attempts to migrate to regions of lesser load. Because of the relative impermeability, water becomes trapped in the voids in the shale and can only migrate slowly. As the load is increased there comes a point when it is in part transferred to the pore water, resulting in a build-up of pore pressure. Depending on the permeability of the shale and the rate of loading, the pore pressure can more or less increase in value so that it equals the pressure imposed by the load. This greatly reduces the shear strength of the shale and a serious failure can occur, especially in the weaker compaction shales. For instance, high pore pressure in the Pepper Shale was largely responsible for the foundation failure at Waco Dam, Texas (see Underwood[52]). Pore pressure problems are not so important in cemented shales.

Clay shales usually have permeabilities of the order 1×10^{-8} to 10^{-12} m/s; whereas sandy and silty shales and closely jointed cemented shales may have permeabilities as high as 1×10^{-6} m/s. However, Jumikis[6] noted that where the Brunswick Shale in New Jersey was highly fissured, it could be used as an aquifer. He also noted that the build-up of ground water pressure along joints could cause shale to lift bedding planes and lead to slabs of shale breaking from the surface. Hence in this shale formation subsurface water must be drained by an efficient system to keep excavations dry. The Brunswick Shale is generally covered with a mantle of weathered material which may be up to 1.2 m thick, grading upwards into a residual soil, which also in places may be a metre or so thick. These materials disintegrate rapidly on wetting and drying, and on freezing. In fact the silty residual soil is subject to frost heave and frost boils. The weathered material cannot support foundations for heavy structures.

8.7 MARL

Marl is a term which has been assigned various meanings, although it has recently been defined by Pettijohn[62] as a rock with 35 to 65% carbonate and a complementary content of clay. However, this definition cannot be applied to many of the rocks in Britain which are referred to as marls, for example, most of the marls of the Keuper series contain less than 20% carbonate material. Such rocks according to the classification of clay-lime carbonate mixtures, after Barth, Correns and Eskola[63], are marly clays or mudstones. In the account which follows the term marl is used in the latter context.

Because the marls of the Keuper series are distributed extensively in the UK and because they have been more thoroughly investigated as far as their geotechnical properties are concerned, attention will be focused on them alone.

The marls of the Keuper series consist of between 50 and 90% clay minerals. The marl may contain thin veins or beds of gypsum, in such cases the ground water contains sulphates. According to Dumbleton[64] illite accounts for 28 to

56% of the clay minerals and chlorite may total some 39%. Usually more than half the chlorite is of the swelling type. Quartz tends to vary between 5 and 35%. The other minerals include calcite and dolomite (which usually comprise less than 20%) and hematite -- one or two per cent. Occasionally other clay minerals such as sepiolite and palygorskite have been found.

The clay particles tend to be aggregated mainly into silt-size units, the aggregated structure being extremely variable (see Davis[65]). Particles composing the aggregate are held together by cement. Sherwood[66] suggested that silica might

Table 8.8 REQUIREMENTS FOR CONCRETE EXPOSED TO SULPHATE ATTACK

Concentration of sulphates expressed as SO_3			Minimum cement content			
			Nominal maximum size of aggregate			Maximum free water/ cement ratio
Class	In soil, total SO_3	In ground water	Type of cement	40 mm	20 mm	10 mm	
	%	parts per 100 000		kg/m^3	kg/m^3	kg/m^3	
1	less than 0.2	less than 30	Ordinary Portland or Portland-blastfurnace	240	280	330	0.55
2	0.2–0.5	30–120	Ordinary Portland or Portland-blastfurnace	290	330	280	0.50
			Sulphate-resisting Portland	240	280	330	0.55
			Supersulphated	270	310	360	0.50
3	0.5–1.0	120–250	Sulphate-resisting Portland or supersulphated	290	330	380	0.50
4	1.0–2.0	250–500	Sulphate-resisting Portland or supersulphated	330	370	420	0.45
5	over 2	over 500	As for Class 4, but with the addition of adequate protective coatings of inert material such as asphalt or bituminous emulsions reinforced with fibreglass membranes				

Notes. This table applies only to concrete made with aggregates complying with the requirements of BS 882 or BS 1047 placed in near-neutral groundwaters of pH 6 to pH 9, containing naturally occurring sulphates but not contaminants such as ammonium salts. Concrete prepared from ordinary Portland cement would not be recommended in acidic conditions (pH 6 or less); sulphate-resisting Portland cement is slightly more acid-resistant but no experience of large-scale use in these conditions is currently available. Supersulphated cement has given an acceptable cement, provided that the concrete is dense and prepared with a free water/cement ratio of 0.40 or less in mineral acids down to pH 3.5.

The cement contents given in Class 2 are the minima recommended by the manufacturers. For SO_3 contents near the upper limit of Class 2, cement contents above these minima are advised.

For severe conditions, e.g. thin sections, sections under hydrostatic pressure on one side only and sections partly immersed, consideration should be given to a further reduction of water/cement ratio and, if necessary, an increase in cement content to ensure the degree of workability needed for full compaction and thus minimum permeability.

(*After BRE Digest 174*[60]. *Crown Copyright.* Reproduced by permission of the Controller of HMSO and by courtesy of the Director of the Building Research Establishment, UK)

be the cementing agent whilst Lees[67] assumed that the clay particles were bound together by physical forces. The former explanation seems the more likely. Aggregation leads to the lack of correlation between consistency limits and shear strength on the one hand and clay content on the other. Because engineering behaviour is controlled by the aggregates, rather than the individual clay minerals, the plasticity, according to Davis[68], is lower than would be expected. He consequently proposed the aggregation ratio as a means of assessing the degree of aggregation. The aggregation ratio was defined as the percentage weight of clay as determined by mineralogical analysis, expressed as a ratio of the percentage weight of clay particles determined by sedimentation techniques. On the other hand Sherwood[66] maintained that aggregation did not give anomalous plasticity values for Keuper marls and that they can generally be classified as materials of low to medium plasticity (Table 8.9). The activity of the Keuper marls increases as the degree of aggregation increases.

Table 8.9 SOIL CLASSIFICATION TESTS ON KEUPER MARL (From Sherwood)

	Clay content by sedimentation (%)	Clay content by X-ray analysis (%)	Liquid limit (%)	Plastic limit (%)	Plasticity index (%)
1	26	94	71	40	31
2	36	58	33	19	14
3	30	87	46	28	25
4	12	77	48	29	19

These marls are very often fissured, weathering and water penetrating the fissures and thereby further reducing the strength of the material. Certain marls exhibit rapid softening when exposed to wet conditions. The fissures close with increasing depth. Skempton and Davis (see Chandler[69]) proposed the following classification of weathered Keuper marl:

Zone	Description
V Fully weathered	Matrix only
IV Highly weathered	Matrix with occasional pellets < 3 mm diameter
III Moderately weathered	Matrix with frequent lithorelicts up to 25 mm
II Slightly weathered	Angular blocks of unweathered marl with virtually no matrix
I Unweathered	Marl (often fissured)

Weathering first develops along fissures in zone II material, the weathered product consisting of a thin veneer of silt. A significant proportion of the marl is weathered in zone III, the unweathered material occurring as angular fragments set in a weathered matrix which is predominantly silty. The water content of the matrix exceeds that of the lithorelicts. In zone IV the lithorelicts are mainly of coarse sand size, and the marl has lost much of its silty texture, indeed up to 50% may be composed of clay-sized particles. This indicates that particle aggregation is broken down upon weathering. Little or no trace of the original structure now remains and the material has a lower permeability than has zone III, 5×10^{-9} to 5×10^{-10} m/s as compared with 1×10^{-8} to 1×10^{-9} m/s respectively. Finally the marl is completely weathered, becoming a plastic, slightly silty, clay. Chandler[69] showed that highly and fully weathered marl could be distinguished from material from zones I, II and III by their particle size distribution and plasticity index (Table 8.10).

An extensive review of the Keuper Marl as a foundation material has been provided by Meigh[70].

Table 8.10 SOME INDEX PROPERTIES OF KEUPER MARL (After Chandler)

Index property	Weathering zone		
	I and II	III	IV
Bulk density, Mg/m³	2.5−2.3	2.3−2.1	2.2−1.8
Dry density, Mg/m³	2.4−1.9	2.1−1.8	1.8−1.4
Natural moisture content (%)	5−15	12−20	18−35
Liquid limit (%)	25−35	25−40	35−60
Plastic limit (%)	17−25	17−27	17−33
Plasticity index	10−15	10−18	17−35
% clay size (BS 1377)	10−35	10−35	30−50
Aggregation ratio (A_r)	10−2.5	10−2.5	2.5
c' (kPa)	⩾ 27.6	⩽ 17.2	⩽ 17.2
ϕ'	40°	42°−32°	32°−25°
ϕ'_r	32°−23°	29°−22°	24°−18°
$\dfrac{\tau_{max} - \tau_{res}*}{\tau_{max}}$ (%)	55	55−30	35−20

*Percentage reduction from peak to residual strength

8.9 SEATEARTHS AND COAL SEAMS

Seatearths are almost invariably found beneath coal seams. Indeed they represent fossil soils and as such are characterised by the presence of fossil rootlets. These tend to destroy the lamination and bedding. The character of a seatearth depends on the type of deposit which was laid down immediately before the establishment of plant growth. If this was mud then a fireclay underlies the coal, whereas if silts and sands were deposited then a gannister was subsequently formed. Many gannisters are pure siltstones and because they are usually well cemented they are hard and strong. Fireclays with a low quartz content are typically highly slickensided and break easily along randomly orientated listric surfaces. The presence of listric surfaces may mean that a fireclay will disintegrate within a few wetting and drying cycles.

Most coal seams are composite in character. At the base the coal is often softer and is sometimes simply referred to as 'bottom coal'. Bright coal is often of most importance in the centre of a seam whilst dull coal may predominate in the upper part of a seam. Coal generally breaks into blocks which have three faces approximately perpendicular to each other. These surfaces are referred to as cleat. The cleat direction is usually fairly constant and is best developed in bright coal. Cleat perhaps may be coated with films of mineral matter, commonly calcite, ankerite and pyrite. Coal seams may split or be replaced, totally or partially, by washouts. Coal is more suspect to mechanical than chemical weathering. Its crushing strength varies but generally it is less than 20 MPa (Table 8.6 and Bell[71]).

References

1. Terzaghi, K., 'Introduction to tunnel geology.' In *Rock Tunnelling with Steel Supports* by R. Proctor and T. White Commercial Shearing and Stamping Co., Youngstown, Ohio, 17–99 (1946).
2. Terzaghi, K., 'Dam foundations on sheeted rock.' *Geotechnique,* 12, 199–208 (1962).
3. Jaeger, C., 'The Malpasset report,' *Water Power,* 15, 55–61 (1963).
4. Jaeger, C., 'The stability of partly immersed fissured rock masses and the Vajont Slide,' *Civ. Engng. Pub. Works Rev.,* 64, 1204–1207 (1969).
5. Pettijohn, P.J., Potter, P.E. and Siever, R., *Sands and sandstones,* Springer-Verlag, Berlin (1975).
6. Bell, F.G., 'The physical and mechanical properties of the Fell Sandstone,' *Engng. Geol.,* 12, 1-29 (1978).
7. Price, N.J., 'The compressive strength of Coal Measures rocks,' *Colliery Guardian,* 283–292 (1960).
8. Price, N.J., 'The influence of geological factors on the strength of Coal Measures rocks,' *Geol. Mag.,* 100, 428–443 (1963).
9. Moore, J.F.A., 'A long term plate test on Bunter Sandstone,' *Proc. 3rd Int. Cong. Rock. Mech., Denver,* 2, 724–732 (1974).
10. Moore, J.F.A. and Jones, C.W., '*In situ* deformation of Bunter Sandstone'. In *Settlement of Structures,* British Geotechnical Society, Pentech Press, London, 311–319 (1975).
11. Hill, H.P., 'The Ladybower Reservoir,' *J. Inst. Water Engrs.,* 3, 414–433 (1949).
12. Fookes, P.G. and Higginbottom, I.E., 'The classification and description of near-shore carbonate sediments for engineering purposes,' *Geotechnique,* 25, 406–411 (1975).
13. Sowers, G.F., 'Failures in limestones in the humid subtropics,' *Proc. ASCE, J. Geot. Engng. Div.* GT8, 101, 771–87 (1975).
14. Kennard, M.F. and Knill, J.L., 'Reservoirs on limestone, with particular reference to the Cow Green scheme,' *J. Inst. Water Engrs.,* 23, 87–113 (1969).
15. Frink, J.W., 'The foundations of Hales Bar Dam,' *Econ. Geol.* 41, 576–92 (1946).
16. North, F.J., Some geological aspects of subsidence not due to mining.' *Proc. S. Wales Inst. Engrs.,* 52, 127–58 (1951).
17. Jennings, J.E., Brinto, A.B.A., Louw, A. and Gowan, G.D., 'Sinkholes and subsidences in the Transvaal dolomite of South Africa,' *Proc. 6th Conf. Soil Mech. Found. Engng., Montreal,* 1, 51–4 (1965).
18. Bell, F.G., 'A note on the geotechnical properties of chalk,' *Engng. Geol.,* 11, 221–6 (1977).
19. Higginbottom, I.E., 'The engineering geology of the Chalk,' *Proc. Symp. on Chalk in Earthworks, I.C.E.,* London, 1–14 (1965).
20. Carter, P.G. and Mallard, D.J., 'A study of the strength, compressibility and density trends within the Chalk of South East England,' *Q. J. Engng. Geol.,* 7, 43–56 (1974).
21. Meigh, A.C., and Early, K.R., 'Some physical and engineering properties of chalk,' *Proc. 4th Int. Conf. Soil Mech. Found. Engng.,* 1, 68–73 (1957).
22. Wakeling, T.R.M., 'Foundations on chalk,' *Proc. Symp. on Chalk in Earthworks, I.C.E.,* London, 15–23 (1965).
23. Anon. Engineering Group of Geological Society Working Party Report: 'The logging of cores for engineering purposes,' *Q. J. Engng. Geol.,* 3, 1–24 (1970).
24. Ward, W.H., Burland, J.B. and Gallois, R.W., 'Geotechnical assessment of a site at Mundford, Norfolk for a large Proton accelerator,' *Geotechnique,* 18, 399–431 (1968).
25. Wakeling, T.R.M., 'A comparison of the results of standard site investigation methods against the results of a detailed geotechnical investigation in Middle Chalk at Mundford, Norfolk.' In *In situ investigations in soils and rocks,* British Geotechnical Society, London, 17–22 (1970).
26. Dennehy, J.P., 'Correlating the SPT N value with chalk grade from some zones of the Upper Chalk,' *Geotechnique,* 26, 610–614 (1976).
27. Burland, J.B. and Lord, J.A., 'The load deformation behaviour of Middle Chalk at Mundford, Norfolk: a comparison between full-scale performance and *in situ* and laboratory measurements.' In *In situ investigations in soils and rocks,* British Geotechnical Society, London, 3–16 (1969).

28. Burland, J.B., Kee, R. and Burford, D., 'Short term settlement of a five-storey building on soft chalk.' In *Settlement of structures,* British Geotechnical Society, Pentech Press, London, 259–265 (1974).
29. Kee, R., Parker, A.S. and Wehale, J.E.C., 'Settlement of a twelve-storey building on piled foundations in chalk at Basingstoke.' In *Settlement of structures,* British Geotechnical Society, Pentech Press, London, 275–282 (1974).
30. Ineson, J., 'A hydrogeological study of the permeability of chalk,' *J. Inst. Water Engrs.,* 16, 255–286 (1962).
31. West, G. and Dumbleton, M.J., 'Some observations on swallow holes and mines in the Chalk,' *Q. J. Engng. Geol.,* 5, 171–178 (1972).
32. Lewis, W.A. and Croney, D., 'The properties of chalk in relation to road foundations and pavements,' *Proc. Symp. on Chalk in Earthworks, I.C.E.,* London, 27–42 (1965).
33. Justo, J.L. and Zapico, L., 'Compression between measured and estimated settlements at two Spanish aqueducts on gypsum rock.' In *Settlement of structures,* British Geotechnical Society, Pentech Press, London, 266–274 (1975).
34. Eck, W. and Redfield, R.C., 'Engineering geology problems at Sanford Dam, Texas,' *Bull. Ass. Engng. Geologists,* 3, 15–25 (1965).
35. Brune, G., 'Anhydrite and gypsum problems in engineering geology,' *Bull Ass. Engng. Geologists,* 3, 26–38 (1965).
36. Kendal, P.F. and Wroot, H.E., 'The geology of Yorkshire,' *Printed privately* (1924).
37. Redfield, R.C., 'Brantley Reservoir site – an investigation of evaporite and carbonate facies,' *Bull Ass. Engng. Geologists,* 6, 14–30 (1968).
38. James, A.N. and Lupton, A.R.R., 'Gypsum and anhydrite in foundations of hydraulic structures,' *Geotechnique,* 28, 249–273 (1978).
39. Hawkins, A.B., 'Case histories of some effects of solution/dissolution in the Keuper rocks of the Severn estuary,' *Q. J. Engng. Geol.,* 12, 31–40 (1979).
40. Johnson, W.D., 'The high plains and their utilization,' *U.S Geol. Surv., 21st Annual Rept.,* Part 4, 601–741 (1901).
41. Bell, F.G., 'Salt and subsidence in Cheshire, England,' *Engng. Geol.,* 9, 237–247 (1975).
42. Kuenen, Ph. H., 'Geotechnical calculations concerning the total mass of sediments of the Earth,' *Am. J. Sci.,* 239, 161–190 (1941).
43. Ingram, R.L., 'Fissility of Mudrocks,' *Bull Geol. Soc. Am.,* 64, 869–78 (1953).
44. Mead, W.J., 'Engineering geology of dam sites,' *Trans. 2nd Int. Congr. Large Dams, Washington, D.C.,* 4, 183–198 (1936).
45. Kennard, M.F., Knill, J.L. and Vaughan, P.R., 'The geotechnical properties and behaviour of carboniferous shale at Balderhead Dam,' *Q. J. Engng. Geol.,* 1, 3–24 (1967).
46. De Graft-Johnson, J.W.S., Bhatia, H.S. and Yeboa, S.L., 'Geotechnical properties of Accra Shales,' *Proc. 8th Int. Conf. Soil Mech. Found. Engng., Moscow,* 2, 97–104 (1973).
47. Morgenstern, N.R. and Eigenbrod, M.D., 'Classification of argillaceous soil and rocks,' *Proc. ASCE, J. Geot. Engng. Div, GT10,* 100, 1137–56 (1974).
48. Taylor, R.K. and Spears, D.A., 'The breakdown of British Coal Measures rocks, *Int. J. Rock Mech. Min. Sci.,* 7, 481–501 (1970).
49. Badger, C.W., Cummings, A.D. and Whitmore, R.L., 'The disintegration of shale,' *J. Inst. Fuel,* 29, 417–23 (1956).
50. Grice, R.H., 'Test procedures for the susceptibility of shale to weathering,' *Proc. 7th Int. Conf. Soil Mech. Found. Engng., Mexico City,* 3, 884–889 (1969).
51. Casagrande, A., *Notes on swelling characteristics of clay-shales,* Harvard Soil Mechanics Series, Harvard Univ., Cambridge, Mass. (1949).
52. Underwood, L.B., 'Classification and identification of shales,' *Proc. ASCE Soil Mech. Found. Eng. Div.,* 93, No. SM6, 97–116 (1967).
53. Zaruba, Q. and Bukovansky, M., 'Mechanical properties of Ordovician shales of central Bohemia,' *Proc. 6th Int. Conf. Soil Mech. Found. Engng, Montreal,* 3, 421–4 (1965).
54. Chappell, B.A., 'Deformational response of differently-shaped and sized test pieces of shale rock,' *Int. J. Rock Mech. Min. Sci.,* 11, 21–28 (1974).
55. Burwell, E.B., 'Geology in dam construction, Part I.' In *Applications of geology to engineering practice, Berkey Volume* Geol. Soc. Am., 11-31 (1950).
56. Smith, C.K. and Redlinger, J.F., 'Soil properties of the Fort Union clay shale,' *Proc. 3rd Int. Conf. Soil Mech. Found. Engng. Zurich,* 1, 56–61 (1953).

57. Underwood, L.E., Thorfinnson, S.T. and Black, W.T., 'Rebound in redesign of the Oake Dam hydraulic structures,' *Proc. ASCE, J. Soil Mech. Found. Engng. Div.,* SM2, 90, 3830, 859–68 (1964).

58. Fasiska, E., Wagenblast, N. and Dougherty, M.T., 'The oxidation mechanism of sulphide minerals,' *Bull. Ass. Engng. Geologists,* 11, No. 1, 75–82 (1974).

59. Penner, E., Eden, W.J. and Gillott, J.E., 'Floor heave due to biochemical weathering of shale,' *Proc. 8th Int. Conf. Soil Mech. Found. Engng. Moscow,* 2, 151–158 (1973).

60. BRS Digest 90, *Concrete in sulphate bearing soils and ground waters,* Building Research Establishment, Garston, Watford, HMSO (1975).

61. Jumikis, A.R., 'Some engineering aspects of the Brunswick shale,' *Proc. 6th Int. Conf. Soil Mech. Found. Engng, Montreal,* 2, 99–102 (1965).

62. Pettijohn, F.J., *Sedimentary rocks,* Harper & Row, New York (1975).

63. Barth, T.F.W., Correns, C.W. and Eskola, P., *Die enstchung der gestiene,* Springer, Berlin (1939).

64. Dumbleton, M.J., 'Origin and mineralogy of African red clays and Keuper marl,' *Q. J. Engng. Geol.,* 1, 39–46 (1967).

65. Davis, A.G., 'The structure of Keuper Marl.,' *Q. J. Engng. Geol.,* 1, 145–153 (1968).

66. Sherwood, P.T., 'Classification tests on African red clays and Keuper marl,' *Q. J. Engng. Geol.,* 1, 47–56 (1967).

67. Lees, G., 'Geology of the Keuper Marl,' *Proc. Geol. Soc. Lond.,* No. 1621, 46 (1965).

68. Davis, A.G., 'On the mineralogy and phase equilibrium of Keuper marl,' *Q. J. Engng. Geol.,* 1, 25–46 (1967).

69. Chandler, R.J., 'The effects of weathering on the shear strength properties of Keuper marl,' *Geotechnique,* 19, 321–334.

70. Meigh, A.C., 'The Triassic rocks, with particular reference to predicted and observed performance of some major foundations,' *Geotechnique,* 26, 391–452 (1976).

71. Bell, F.G., 'The Character of the Coal Measures.' In *Site Investigations in Areas of Mining Subsidence,* ed. F.G. Bell, Butterworths, London, 25–40 (1975).

Chapter 9

Subsurface Water and Ground Conditions

The principal source of ground water is precipitation or meteoric water. The amount of water which percolates into the ground depends upon how that precipitation is dispersed, that is, on what proportions are assigned to immediate run-off and to evapotranspiration, the remainder constituting the proportion allotted to percolation. Infiltration is the movement of surface water into the ground, percolation being its subsequent movement to the zone of saturation, but in reality one cannot be separated from the other.

9.1 CAPILLARY ACTION

The retention of water in a soil depends upon the capillary force and the molecular attraction of the particles. When the pores in a soil become thoroughly wetted the capillary force declines so that gravity becomes more effective. In this way downward percolation can continue after infiltration has ceased, but as the soil dries capillarity increases in importance. No further percolation occurs after the capillarity and gravity forces are balanced. Thus water percolates into the zone of saturation when the retention capacity is satisfied. The pores within the zone of saturation are filled with water, frequently referred to as phreatic water. The upper surface of this zone is therefore known as the phreatic surface but is more commonly termed the water table. Above the zone of saturation is the zone of aeration in which both air and water occupy the pores.

Capillary movement in a soil refers to the movement of moisture through the minute pores between the soil particles which act as capillaries. It takes place as a consequence of surface tension, therefore moisture can rise from the water table. This movement, however, can occur in any direction, not just vertically. It occurs whenever evaporation takes place from the surface of the soil, thus exerting a 'surface tension pull' on the moisture, the forces of surface tension increasing as evaporation proceeds. Accordingly capillary moisture is in hydraulic continuity with the water table and is raised against the force of gravity, the degree of saturation decreasing from the water table upwards. Equilibrium is attained when the forces of gravity and surface tension are balanced.

The boundary separating capillary moisture from the gravitational water in the zone of saturation is, as would be expected, ill defined and cannot be determined accurately. That zone immediately above the water table which is saturated with capillary moisture is referred to as the closed capillary fringe, whilst

Table 9.1 CAPILLARY RISES AND PRESSURES IN SOILS (After Jumikis[1])

Soil	Capillary rise (mm)	Capillary pressure (kPa)
Fine gravel	Up to 100	Up to 1.0
Coarse sand	100–150	1.0–1.5
Medium sand	150–300	1.5–3.0
Fine sand	300–1000	3.0–10.0
Silt	1000–10000	10.0–100.0
Clay	Over 10000	Over 100.0

above this air and capillary moisture exist together in the pores of the open capillary fringe. The depth of the capillary fringe is largely dependent upon the particle size distribution and density of the soil mass, which in turn influence pore size. In other words the smaller the pore size, the greater is the depth. For example, capillary moisture can rise to great heights in clay soils (Table 9.1) but the movement is very slow. In soils which are poorly graded the height of the capillary fringe generally varies, whereas in uniformly textured soils it attains roughly the same height.

Where the water table is at shallow depth and the maximum capillary rise is large, moisture is continually attracted from the water table due to evaporation from the surface so that the uppermost soil is near saturation. For instance, under normal conditions peat deposits may be assumed to be within the zone of capillary saturation. In other words, the height to which water can rise in peat by capillary action is greater than the depth below ground to which the water table is reduced by drainage. The coarse fibrous type of peat, containing appreciable sphagnum, may be an exception.

Drainage of capillary moisture cannot be effected by the installation of a drainage system within the capillary fringe as only that moisture in excess of that retained by surface tension can be removed, but it can be lowered by lowering the water table. The capillary ascent, however, can be interrupted by the installation of impermeable membranes or layers of coarse aggregate. These two methods can be used in the construction of embankments, or more simply the height of the fill can be raised.

At each point where moisture menisci are in contact with soil particles the forces of surface tension are responsible for the development of capillary or suction pressure (Table 9.1). This tends to force the soil particles together and these compressive stresses contribute towards the strength and stability of the soil. There is a particular suction pressure for a particular moisture content in a given soil, the magnitude of which is governed by whether it is becoming wetter or drier. In fact as a clay soil dries out the soil suction may increase to the order of several thousands of kilopascals. However, the strength of a soil attributable to soil suction is only temporary and may be destroyed upon saturation.

Below the water table the water contained in the pores is under normal hydrostatic load, the pressure increasing with depth. Because these pressures exceed atmospheric pressure they are designated positive pressures. On the other hand the pressures existing in the capillary zone are less than atmospheric and so are termed negative pressures. Thus the water table is usually regarded as a datum of zero pressure between the positive pore pressure below and the negative above.

9.2 POROSITY AND PERMEABILITY

Porosity and permeability are the two most important factors governing the accumulation, migration and distribution of water in soils and rocks. However, porosity and permeability are not necessarily as closely related as would be expected, for instance, very fine textured sandstones frequently have a higher porosity than coarser ones though the latter are more permeable. In other words the size of the pores is all important as far as the permeability of a formation is concerned. It is not uncommon to find variations in both porosity and permeability throughout a formation. The factors affecting the porosity of a rock include particle size distribution, grain shape, fabric, degree of compaction and cementation, solution effects, and lastly mineralogical composition, particularly the presence of clay particles. The highest porosity is commonly obtained when the grains are all of the same size.

In ordinary hydraulic usage a substance is termed permeable when it permits the passage of a measurable quantity of fluid in a finite period of time and impermeable when the rate at which it transmits that fluid is slow enough to be

Table 9.2 VALUES OF SOIL PERMEABILITIES

Degree of permeability	Range of coefficient of permeability (m/s)	Soil type
High	10^{-3}	Medium and coarse gravel
Medium	10^{-3} to 10^{-5}	Fine gravel; coarse, medium and fine sand; dune sand; clean sand-gravel mixtures
Low	10^{-5} to 10^{-7}	Very fine sand, silty sand, loose silt, loess, well fissured clays
Very low	10^{-7} to 10^{-9}	Dense silt, dense loess clayey silt, poorly fissured clays
Impermeable	10^{-9}	Unfissured clays

negligible under existing temperature-pressure conditions (Table 9.2). The flow through a unit cross section of material is modified by temperature, hydraulic gradient and the coefficient of permeability. The latter is affected by the range of grain size and shape, stratification, consolidation and cementation of the material. Temperature changes affect the flow rate of a fluid by changing its viscosity. The rate of flow is commonly assumed to be directly proportional to the hydraulic gradient but this is not always so in practice.

Determination of the permeability of a sample in the laboratory does not necessarily bear any relation to the permeability of the formation concerned in the field. This is because most soil masses are anisotropic in character and in rock masses the discontinuities are the most important conduits for water movement (see Ineson[2], Bell[3]). Discontinuities allow water to percolate through rocks with extremely low values of porosity. Indeed the frictional resistance to flow through discontinuities is frequently much lower than that offered by a porous medium, hence appreciable quantities of water may be transmitted. However, discontinuities tend to close with depth. Similarly the permeability of soil, even if fairly uniform in character, tends to decrease with increasing depth because the increasing weight of overburden leads to its densification. Field tests therefore provide much more reliable results (see Bell[4]). In a sequence of layered strata the permeability of the individual beds will vary, usually the average permeability in the vertical direction will be less than that in the horizontal.

Subsurface water is normally under pressure which increases with increasing depth below the water table to very high values. Such water pressures have a significant influence on the engineering behaviour of most rock and soil masses and their variations are responsible for changes in the stresses in these masses which affect their deformation characteristics and failure. The influence of interstitial water on rock has been briefly mentioned above, a fuller account is given by Serafim[5].

9.3 PORE PRESSURES

Piezometers are installed in the ground in order to monitor and obtain accurate measurements of pore water pressures. Observations should be made regularly so that changes due to external factors such as excessive precipitation, tides, the seasons, etc are noted, it being most important to record the maximum pressures whicn have occurred. Standpipe piezometers allow the determination of the position of the water table and the permeability (see Sherrell[6]). Hydraulic piezometers connected to a manometer board record the changes in pore water pressure. Usually simpler types of piezometer are used in the more permeable soils. The response to piezometers in rock masses can be very much influenced by the incidence and geometry of the discontinuities so that the values of water pressure obtained may be misleading if due regard is not given to these structures. The efficiency of a soil in supporting a structure is influenced by effective or intergranular pressure, that is, the pressure between the particles of the soil which develops resistance to applied load. Because the moisture in the pores offers no resistance to shear it is ineffective or neutral and therefore pore water pressure has also been referred to as neutral pressure. Since the pore or neutral pressure plus the effective pressure equals the total pressure, reduction in pore pressure increases the effective pressure. Reduction of the pore pressure by drainage consequently affords better conditions for carrying a proposed structure. The effective pressure at a particular depth is simply obtained by multiplying the unit weight of the soil by the depth in question and subtracting the pore pressure for that depth.

In a layered sequence the individual layers may have different unit weights. The unit weight of each should then be multiplied by its respective thickness and the pore pressure involved subtracted. The effective pressure for the total thickness concerned is then obtained by summing the effective pressures of the individual layers. Water held in the capillary fringe by soil suction does not affect the values of pore pressure below the water table. However, the weight of water held in the capillary fringe does increase the weight of overburden and so the effective pressure.

Volume changes brought about by loading compressive soils depend upon the level of effective stress and are not affected by the area of contact. The latter may also be neglected in saturated or near saturated soils.

There is some evidence which suggests that the law of effective stresses as used in soil mechanics, in which the pore pressure is subtracted from all direct stress components, holds true for some rocks, those with low porosity may at times prove the exception. However, Serafim[5] suggested that it appeared that pore pressures have no influence in brittle rocks. This is probably because the strength of such rocks is mainly attributable to the strength of the bonds between the component crystals or grains.

9.4 COEFFICIENT OF PERMEABILITY

If at two different points within a continuous mass of water there are different amounts of energy, then flow takes place towards the point of lesser energy and the difference in head is expended in maintaining that flow. Other things being equal the velocity of flow between two points is directly proportional to the difference in head between them. The hydraulic gradient refers to the loss of head or energy of water flowing through the ground. This loss of energy by the water is due to the friction resistance of the ground material, and this is greater in fine than coarse grained soils. Thus, in a given engineering project, there is no guarantee that the rate of flow will be uniform and indeed this is the exception. However, if it is assumed that the resistance to flow is constant, then for a given difference in head the flow velocity is directly proportional to the flow path.

The permeability of a particular material is defined by its coefficient of permeability (k). The basic law concerned with flow is that enunciated by Darcy[7] which states that the rate of flow (v) per unit area is proportional to the gradient of the potential head (i) measured in the direction of flow:

$$v = ki$$

and for a particular rock or soil or part of it, of area (A):

$$Q = vA = Aki$$

where Q is the quantity in a given time. The ratio of the cross sectional area of the pore spaces in a soil to that of the whole soil is given by $e/(1 + e)$, where e is the void ratio. Hence a truer velocity of flow, that is, the seepage velocity (v_s) is:

$$v_s = \left(\frac{1 + e}{e}\right) ki$$

Darcy's law is valid as long as a laminar flow exists; departures from this law therefore occur when the flow is turbulent. They also occur when the velocity of flow is high. Such conditions exist in very permeable media, normally when the Reynolds number can attain values above four. In other words, it is usually accepted that this law can be applied to those soils which have finer textures than gravels. Furthermore Darcy's law probably does not accurately represent the flow of water through a porous medium of extremely low permeability, because of the influence of surface and ionic phenomena and the presence of gases.

Darcy omitted to recognise that permeability also depends upon the density (ρ) and viscosity of the fluid(μ) and the average pore size of the porous medium (d) as well as the pore shape. In fact permeability is directly proportional to the unit weight of the fluid concerned and is inversely proportional to its viscosity.

The latter is very much influenced by temperature. The hydraulic conductivity (k_c) is an attempt to take these factors into account and is expressed as follows:

$$k_c = Cd^2 \frac{\rho}{\mu}$$

where C is a dimensionless constant or shape factor which takes note of the effects of stratification, packing, particle size distribution and porosity.

It is assumed in the above expression that both the porous medium and the water are mechanically and physically stable, but this may never be true. For example, ion exchange on clay and colloid surfaces may bring about changes in mineral volume which in turn affect the shape and size of the pores. Moderate to high ground water velocities tend to move colloids and clay particles. Solution and deposition may occur from the pore fluids. Small changes in temperature and/or pressure may cause gas to come out of solution which may block pore spaces.

Generally it is the interconnected systems of discontinuities which determine the permeability of a particular rock mass. Indeed the permeability of a jointed rock mass is usually several orders higher than that of intact rock. According to Serafim[5] the following expression can be used to derive the filtration through a rock mass intersected by a system of parallel sided joints with a given opening (e) separated by a given distance (d):

$$k = \frac{e^3 \gamma_w}{12d \, \mu}$$

where γ_w is the unit weight of water. The velocity of flow through a single joint of constant gape is expressed by:

$$v = \left(\frac{e^2 \gamma_w}{12 \, \mu} \right) i$$

where i is the hydraulic gradient.

Subsequently Wittke[8] suggested that where the spacing between discontinuities is small in comparison with the dimensions of the rock mass it is often admissible to replace the fissured rock, with regard to its permeability, by a continuous anisotropic medium. The permeability of this can be described by means of Darcy's law. Wittke also provided a resume of procedures by which three dimensional problems of flow through rocks under complex boundary conditions could be solved.

Lovelock *et al*[9] suggested that it can be shown that the contribution of the fissures (T_f) to the transmissivity of an idealised acquifer can be approximated from the following expression:

$$T_f = \frac{g}{12\mu_k} \sum_{x=1}^{n} b_x^{\,3}$$

where b_x is the effective aperture of the xth of n horizontal, parallel-sided smooth-walled openings; g is the acceleration due to gravity; μ_k is the kinematic viscosity of the fluid, and flow is laminar. The third power relationship means that a small variation in effective aperture (b) gives rise to a large variation in the fissure contribution (T_f). A fuller account of the theory of ground water movement can be obtained from De Weist[10].

9.5 FLOW NETS

Flow nets provide a graphical representation of the flow of water through the ground and indicate the loss of head involved. They also provide data relating to the changes in head velocity and effective pressure which occur in a foundation subjected to flowing ground water conditions. For instance, when the flow lines of a flow net move closer together this indicates that the flow increases, although their principal function is to indicate the direction of flow. The equipotential lines indicate equal losses in head or energy as the water flows through

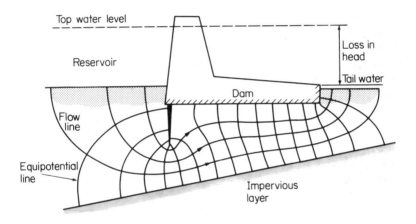

Figure 9.1 Flow net beneath concrete gravity dam, with cut-off at the heel, showing seventeen equipotential drops and four flow channels

the ground, so that the closer they are the more rapid is the loss in head. Hence a flow net can provide quantitative data related to the flow problem in question, for example, seepage pressures can be determined at individual points within the net (Figure 9.1).

9.6 CRITICAL GRADIENT AND QUICK CONDITIONS

As water flows through the soil and loses head, its energy is transferred to the particles past which it is moving, which in turn creates a drag effect on the particles. If the drag effect is in the same direction as is the force of gravity, then the effective pressure is increased and the soil is stable. Indeed the soil tends to become more dense. Conversely if water flows towards the surface then the drag effect is counter to gravity thereby reducing the effective pressure between particles. If the velocity of upward flow is sufficient it can buoy up the particles so that the effective pressure is reduced to zero. This represents a critical condition where the weight of the submerged soil is balanced by the upward acting seepage force. The critical hydraulic gradient (i_c) involved can be calculated from the following expression:

$$i_c = \frac{G_s - 1}{1 + e}$$

where G_s is the relative density of the particles and e is the void ratio.

A critical condition sometimes occurs in loosely packed, saturated silts and sands; if the upward velocity of flow increases beyond the critical hydraulic gradient a quick condition develops.

Quicksands, if subjected to deformation or disturbance, can suffer a spontaneous loss of shear strength. This loss of strength causes them to flow like viscous liquids. Terzaghi[11] explained the quicksand phenomenon in the following terms. Firstly, the sand must be saturated and loosely packed. Secondly, on disturbance the constituent grains become more closely packed, but this leads to an increase in pore water pressure which reduces the forces acting between the grains. This brings about the reduction in strength. If the pore water can escape very rapidly the loss in strength is momentary. Hence the third condition requires that pore water cannot readily escape. This is fulfilled if the sand has a low permeability and/or the seepage path is long.

Casagrande[12] demonstrated that a critical porosity existed above which a quick condition could be developed. He maintained that many coarse grained sands, even when loosely packed, have porosities approximately equal to the critical condition whilst medium and fine grained sands, particularly if uniformly graded, exist well above the critical porosity when loosely packed. Accordingly fine sands tend to be more unstable than coarse grained varieties. It must also be remembered that their permeability is lower.

Quick conditions brought about by seepage forces are frequently encountered in excavations made in fine sands which are below the water table as, for example, in cofferdam work. As the velocity of the upward seepage force increases further from the critical gradient the soil begins to boil more and more violently. At such a point structures fail by sinking into the quicksand. Liquefaction of potential quicksands may be caused by sudden shocks such as the action of heavy machinery (notably pile driving), blasting and earthquakes. Such shocks increase the stress carried by the water, the neutral stress, and give rise to a decrease in the effective stress and shear strength of the soil. There is also a possibility of a quick condition developing in a layered soil sequence where the individual beds have different permeabilities. Hydraulic conditions are particularly unfavourable where water initially flows through a very permeable horizon with little loss of head which means that flow takes place under a great hydraulic gradient. Hydraulic uplift can occur in clay deposits (see Moore and Longworth[13]).

There are several methods which may be employed to avoid the development of quick conditions. One of the most effective techniques is to prolong the length of the seepage path thereby increasing the frictional losses and so reducing the seepage force. This can be accomplished by placing a clay blanket at the base of an excavation where seepage lines converge. If sheet piling is used in excavation then the depth to which it is sunk determines whether or not quick conditions will develop. Hence it should be sunk to a depth which avoids a potential critical condition occurring at the base level of the excavation.

The hydrostatic head can also be reduced by means of relief wells and seepage can be intercepted by a well point system placed about the excavation. Furthermore a quick condition may be prevented by increasing the downward acting force. This may be brought about by placing a load on the surface of the soil where seepage is discharging. Gravel filter beds may be used for this purpose. Suspect soils can also be densified, treated with stabilizing grouts or frozen.

When water percolates through heterogeneous soil masses it moves preferentially through the most permeable zones and it issues from the ground as springs. Piping refers to the erosive action of some such springs, where sediments are removed by seepage forces so forming subsurface cavities and tunnels. In order that erosion tunnels may form the soil must have some cohesion, the greater the cohesion the wider the tunnel. In fact fine sands and silts are most susceptible to piping failures. Obviously the danger of piping occurs when the hydraulic gradient is high, that is, when there is a rapid loss of head over a short distance. This may be indicated on a flow net by a close network of squares where the flow is upward. As the pipe develops by backward erosion it nears the source of water supply so that eventually the water breaks into and rushes through the pipe. Ultimately the hole, so produced, collapses from lack of support. Piping has been most frequently noted downstream of dams, the reservoir providing the water source, but leaking drains can also give rise to piping.

Subsurface structures should be designed to be stable with regard to the highest ground water level that is likely to occur. Structures below ground water level are acted upon by uplift pressures. If the structure is weak this pressure can break it and, for example, cause a blow-out of a basement floor or collapse of a basement wall. If the structure is strong but light it may be lifted. Uplift can be taken care of by adequate drainage or by resisting the upward seepage force. Continuous drainage blankets are effective but should be designed with filters to function without clogging. The entire weight of the structure can resist uplift if a raft foundation is used. Anchors grouted into bedrock can also provide resistance to uplift.

If water flowing under pressure through the ground is confined between two impermeable horizons then it is termed artesian water. Artesian conditions are commonly developed in synclinal structures where an aquifer is sandwiched between two impermeable layers but outcrops at a higher elevation than the position at which the pressure is measured. Artesian conditions can cause serious trouble in excavations and both the position of the water table and the piezometric pressures should be determined before work commences. Otherwise excavations which extend close to strata under artesian pressure may be damaged severely due to blow-outs taking place in their floors. Slopes may also fail. Indeed such sites may have to be abandoned. Bleeder wells can be used to control artesian pressures (see Wade and Taylor[14]).

9.7 FROST ACTION IN SOIL

Obviously frost action in a soil is influenced by the initial soil temperature, as well as the air temperature, the intensity and duration of the freeze period, the depth of frost penetration, the depth of the water table, and types of ground and exposure cover. If frost penetrates down to the capillary fringe in fine grained soils, especially silts, then, under certain conditions, lenses of ice may be developed. The formation of such ice lenses may, in turn, cause frost heave and frost boil which may lead to the break up of roads, the lifting of structures or the failure of slopes.

Frost action not only causes local heaving but is followed by a loss in the bearing capacity of a soil when thawing takes place. When the thaw occurs the water liberated greatly exceeds that originally present in the soil. Moreover a soil usually thaws out from the top downwards, however, it is not unknown for some thawing to take place from the bottom. As the soil thaws downwards the upper layers become saturated, and since water cannot drain through the frozen soil beneath, they may suffer a complete loss of strength.

Jumikis[19] listed the following factors as necessary for the occurrence of frost heave, namely, capillary saturation at the beginning of and during the freezing of the soil, a plentiful supply of subsoil water and a soil possessing fairly high capillarity together with moderate permeability. However, an important factor governing frost heave appears to be grain size since this influences moisture movement. For example, well sorted soils in which less than 30% of the particles are of silt size are non-frost heaving (see Glossop and Skempton[15]). Moreover frost heave does not occur in clays because of their low permeability. Indeed Taber[16] gave an upper size limit of 0.007 mm, above which, he maintained, layers of ice do not develop, although Casagrande[17] suggested that the particle size critical to heave formation is 0.02 mm. If the quantity of such particles in a soil is less than 1%, no heave is expected, but considerable heaving may take place if this amount is over 3% in non-uniform soils and over 10% in very uniform soils. In silts the moisture of the upward capillary rise and/or film flow, if frost penetrates downward into the capillary fringe, forms ice lenses under prolonged and severe freezing conditions. This is because silt particles are small enough to provide a comparatively high capillary rise.

Croney and Jacobs[18] suggested that under the climatic conditions experienced in Britain well-drained cohesive soils with a plasticity index exceeding 15% could be looked upon as non-frost susceptible. Where the drainage is poor and the water table is within 0.6 m of formation level they suggested that the limiting value of plasticity index should be increased to 20%. In experiments with sand they noted that as the amount of silt added was increased up to 55% or the clay fraction up to 30%, increase in permeability in the freezing front was the overriding factor and heave tended to increase. Beyond these values the decreasing permeability below the freezing zone became dominant and progressively reduced the heave. In other words the permeability below the frozen zone was principally responsible for controlling heave. These two authors also suggested that the permeability of soft chalk is sufficiently high to permit very serious frost heave but in the harder varieties the lower permeabilities minimize or prevent heaving.

Maximum heaving, according to Jumikis[19], does not necessarily occur at the time of maximum depth of the 0°C line, there being a lag between the minimum air temperature prevailing and the maximum penetration of the freeze front. In fact soil freezes at temperatures slightly lower than 0°C. As heaves amounting to 30% of the thickness of the frozen layer have frequently been recorded, moisture, other than that initially present in the frozen layer, must be drawn from below, since water increases in volume by only 9% when frozen.

In fact when a soil freezes there is an upward transfer of heat from the ground water towards the area in which freezing is occurring. The thermal energy, in its turn, initiates an upward migration of moisture within the soil. The moisture in the soil can be translocated upwards either in the vapour or liquid phase or by a combination of both. Moisture diffusion by the vapour phase occurs more readily in soils with larger void spaces than in fine grained soils and if a soil is saturated migration in the vapour phase cannot take place.

In a very dense, closely packed soil where the moisture forms uninterrupted films throughout the soil mass, down to the water table, then, depending upon the texture of the soil, the film transport mechanism becomes more effective than the vapour mechanism. The upward movement of moisture due to the film mechanism, in a freezing soil mass, is slow. Nonetheless a considerable amount of moisture can move upwards as a result of this mechanism during the winter. What is more in the film transport mechanism the water table is linked by the films of moisture to the ice lenses.

Before freezing, soil particles develop films of moisture about them due to capillary action. This moisture is drawn from the water table. As the ice lens grows, the suction pressure it develops exceeds that of the capillary attraction of moisture by the soil particles. Hence moisture moves from the soil to the ice lens. But the capillary force continues to draw moisture from the water table and so the process continues.

The amount of segregated ice in a frozen mass of soil depends largely upon the intensity and rate of freezing. When freezing takes place quickly no layers of ice are visible whereas slow freezing produces visible layers of ice of various thicknesses. Ice segregation in soil also takes place under cyclic freezing and thawing conditions.

The frost heave test allows prediction of frost heave (see Jacobs[20]). Unfortunately this type of test is time consuming and so a rapid freeze test has been developed by Kaplar[21]. Approximate predictions of frost heave have also been based on grain size distribution. However, in a recent discussion of frost heaving, Reed et al[22] noted that such predictions failed to take account of the fact that soils can exist at different states of density and therefore porosity, yet they have the same grain size distribution. Also pore size distribution controls the migration of water in the soil and hence, to a large degree, the mechanism if frost heave. They accordingly derived expressions, based upon pore space, for predicting the amount of frost heave (Y) in mm/day:

$$Y = 581.1\,(X_{3\cdot0}) - 5.46 - 29.46\,(X_{3\cdot0})/(X_0 - X_{0\cdot4})$$

where $X_{3\cdot0}$ = cumulative porosity for pores $> 3.0\ \mu m$ but $< 300\ \mu m$,
X_0 = total cumulative porosity,
$X_{0\cdot4}$ = cumulative porosity for pores $> 0.4\ \mu m$ but $< 300\ \mu m$.

A simpler expression based on pore diameters rather than cumulative porosity, somewhat less accurate, was:

$$Y = 1.694(D_{40}/D_{80}) - 0.3805$$

where D_{40} and D_{80} are the pore diameters whereby 40% and 80% of the pores are larger respectively.

Where there is a likelihood of frost heave occurring it is necessary to estimate the depth of frost penetration (see Jumikis[19]). Once this has been done provision can be made for the installation of adequate insulation or drainage within the soil. Alternatively the amount by which the water table needs to be lowered so that it is not affected by frost penetration can be determined. The base of footings should be placed below the estimated depth of frost penetration, which in the UK is generally at a depth of 0.5 m below ground level.

Frost susceptible soils may be replaced by free draining gravels. The addition of certain chemicals to soil can reduce its capacity for water absorption and so can influence frost susceptibility. For example, Croney and Jacobs[18] noted that the addition of calcium lignosulphate and sodium tripolyphosphate to silty soils were both effective in reducing frost heave. The freezing point of the soil may be lowered by mixing in solutions of calcium chloride or sodium chloride, in concentrations of 0.5 to 3.0% by weight of the soil mixture. The heave of non-cohesive soils containing appreciable quantities of fines can be reduced or prevented by the addition of cement or bituminous binders. Addition of cement both reduces the permeability of a soil mass and gives it sufficient tensile strength to prevent the formation of small ice lenses as the freezing isotherm passes through.

9.8 PERMAFROST

Perennially frozen ground or permafrost is characteristic of tundra environments; in fact permafrost covers 20% of the Earth's land surface. During Pleistocene times it was developed over an even larger area, evidence for its former presence being found in the form of fossil ground-ice wedges, solifluction deposits and stone polygons. The temperature of perennially frozen ground below the depth of seasonal change ranges from slightly less than $0°C$ to $-12°C$. Generally the depth of thaw is less the higher the latitude. It is at a minimum in peat or highly organic sediments and increases in clay, silt and sand to a maximum in gravel where it may extend to 2 m depth.

Because permafrost represents an impervious horizon it means that water is prevented from entering the ground so that soils are commonly supersaturated. Hence solifluction deposits and mud flows are typical, indeed soil material on all slopes greater than $1°$ to $3°$ is on the move in summer. Permafrost, by maintaining saturated or supersaturated conditions in surface soils, aids frost stirring, frost splitting and mass wasting processes. Frost action and gravity movements result in certain characteristic features, for example, involutions, frost boils, hummocks, antiplanation terraces, terracettes and stone stripes (see Black[23]). Annual freezing in permafrost areas brings about changes in surface and ground water movements which may result in the development of frost blisters, ice mounds, icings or even pingos.

According to Muller[24] there are two methods of construction in permafrost, namely, the passive and the active methods. In the former the frozen ground is not disturbed or is provided with additional insulation so that heat from the structure does not bring about thawing in the ground below, thereby reducing its stability. By contrast the ground is thawed prior to construction in the active method and it is either kept thawed or removed and replaced by materials not affected by frost action.

Permafrost is an important characteristic, although it is not essential to the definition of periglacial conditions, the latter referring to conditions under which frost action is the predominant weathering agent. A broad survey of periglacial features in Britain and their engineering significance has been provided by Higginbottom and Fookes[25]. Superficial structural disturbances have several causes. The results of frost shattering are particularly noteworthy in the Chalk (see Ward *et al*[26]). Ice wedges may originate as thermal contraction cracks which are enlarged by the growth of ice and subsequently infilled, and in plan may form part of a polygonal network (see Shotton[27]). Surface cracking may be associated with the formation of stone polygons. Involutions are pockets or tongues of highly disturbed material, generally possessing inferior geotechnical properties, which have been intruded from below into overlying layers due to hydrostatic uplift of water trapped beneath a refreezing surface layer (see West[28]). Frost mounds and pingos are subject to solifluction during their lifetime consequently a depression develops when the ice lens melts. This then may be filled with deposits of saturated, normally consolidated fine sediments, which alternate with layers of peat. Mass movements in former periglacial areas in Britain included cambering and valley bulging, landsliding and mudflows.

References

1. Jumikis, A.R., *Soil mechanics,* Van Nostrand, Princeton, N.J. (1968).
2. Ineson, J., 'A hydrogeological study of the permeability of the Chalk,' *J. Inst. Water Engrs.,* 16, 255–286 (1962).
3. Bell, F.G., 'Some petrographic factors relating to porosity and permeability in the Fell Sandstones of Northumberland,' *Q. J. Engng. Geol.,* 11, 113–26 (1978).
4. Bell, F.G., 'In situ testing and geophysical surveying.' In *Foundation Engineering in Difficult Ground,* ed. F.G. Bell, Butterworths, London, 226–280 (1978).
5. Serafim, J.L., 'Influence of interstitial water on rock masses. In *Rock mechanics in engineering practice,* ed. Stagg, K.G. and Zienkiewicz, O.C., Wiley, London, 55–97 (1968).
6. Sherrell, F.W., 'Engineering geology and ground water,' *Ground Engineering,* 9, No. 4, 21–27 (1976).
7. Darcy, H., *Les Fontaines Publiques de la Ville de Dijon,* Dalmont, Paris (1856).
8. Wittke, W., 'Perculation through fissured rock,' *Bull Int. Ass. Engng. Geologists,* No. 7, 3–28 (1973).
9. Lovelock, P.E.R., Price, N. and Tate, T.K., 'Groundwater conditions in the Penrith Sandstone at Cliburn, Westmoreland,' *J. Inst. Water Engrs.,* 29, 157–174 (1975).
10. De Wiest, R.J.H., *Geohydrology,* Wiley, New York (1967).
11. Terzaghi, K., *Erdbaumechanik auf bodenphysikalischer grundlage,* Deuticke, Vienna (1925).
12. Casagrande, A., 'Characteristics of cohesionless soils affecting the stability of slopes and earth fills,' *J. Boston Soc. Civ. Engrs.,* 23, 3–32 (1936).
13. Moore, J.F.A. and Longworth, T.I., 'Hydraulic uplift of the base of a deep excavation in the Oxford Clay,' *Geotechnique,* 29, 35–46 (1979).
14. Wade, N.H. and Taylor, H., 'Control of artesian pressure by bleeder wells,' *Can. Geotech. J.,* 16, 488–496 (1979).
15. Glossop, R. and Skempton, A.W., 'Particle size in silts and sands,' *J. Inst. Civ. Engrs.,* 25, Paper 5492, 81–105 (1945).
16. Taber, S., 'Mechanics of frost heaving,' *J. Geol.,* 38, 303–317 (1930).
17. Casagrande, A., 'Discussion on frost heaving,' *Proc. Highways Res. Board,* 12, 169, Washington D.C. (1932).
18. Croney, D. and Jacobs, J.C., 'The frost susceptibility of soils and road materials,' *Trans. Road Res. Lab., Rept. LR90,* Crowthorne (1967).
19. Jumikis, A.R., 'The soil freezing experiment, *Highway Res. Board Bull. No. 135,*

Factors influencing ground freezing,' *Nat. Acad. Sci.,* Nat. Res. Council Pub. 425, Washington D.C. (1956).

20. Jacobs, J.C., *The Road Research Laboratory frost heave test,* T.R.R.L. Lab. Note LN/766/JCJ (1965).

21. Kaplar, C.W., 'Experiments to simplify frost susceptibility testing of soils,' *US Army Corps of Engrs., Cold Regions Res. and Engng. Lab., Hanover, NH,* Tech. Rept. 223 (1971).

22. Reed, M.A., Lovell, C.W., Altschaeffl, A.G. and Wood, L.E., 'Frost heaving rate predicted from pore size distribution,' *Can. Geotech. J.* **16**, 463–472 (1979).

23. Black, R.F., 'Permafrost'. In *Applied Sedimentation,* ed. Trask, P.D., Wiley, New York, 247–275 (1950).

24. Muller, S.W., 'Permafrost or permanently frozen ground and related engineering problems,' *U.S. Geol. Surv., Spec. Rept., Strategic Engineering Study 62,* Military Intelligence Division Office, Chief of Engineers, U.S. Army (1945). Also published by Edward Bros., Ann Arbor, Mich. (1947).

25. Higginbottom, I.E. and Fookes, P.G., 'Engineering aspects of periglacial features in Britain,' *Q. J. Engng. Geol.,* **3**, 85–118 (1970).

26. Ward, W.H., Burland, J.B. and Gallois, R.W., 'Geotechnical assessment of a site at Mundford, Norfolk, for a large proton accelerator,' *Geotechnique,* **18**, 399–431 (1968).

27. Shotton, F.W., 'Large scale patterned ground in the valley of the Worcestershire Avon,' *Geol. Mag.,* **97**, 407–408 (1960).

28. West, R.G., *Pleistocene geology and biology,* Longmans, London (1968).

Index